Technical Physics

Using the Systems Approach

Text
Student Exercises
Laboratory Activities

J.P. Levasseur

Contents

Laboratory Activities

Acknowledgements

The first edition of this textbook and laboratory manual was used for the first time during the spring semester of 2015. Edition 2 soon followed. It is with great appreciation that my several divisions of physics students routinely searched for and kept track of errors and other format issues in these first editions. A work such as this seems to go on endlessly because improvements can always be made. This edition contains only the errata generated by my fall and spring students of 2016/17. Of note are students Jordon Buhro and Loren Gordon. Many thanks to both these men as well as all of my students during that academic year.

Also, many thanks to my superiors and many coworkers for their unwavering support.

Welcome! Since you're reading this, you are most likely enrolled in a college two-year trade and technical program with four credits in physics required. Your technology-based major program might be computer electronics, wind power technology, automotive technology, precision metals fabrication, or maybe computer aided drafting. All of these fields require an understanding of the fundamental concepts and conventions upon which technologies are applied, developed, controlled, and maintained. More than ever before, rapid changes in technologies are changing the way in which we live, and more and more is expected of our technicians who must adapt and evolve as technologies so quickly advance. By examining the language, writing conventions, standards of measurement, and principles that are the underlying foundation of all technologies, this course is designed to help you gain what you need in order to read, write, compute, and speak tech, and thereby keep up with ever changing technologies. Ultimately, it is hoped you will acquire some of the tools technicians need in order to have a successful, long-term career in a variety of technology-related professions.

Learning can be an incredibly enjoyable and satisfying experience. But be conscious of the fact that information and knowledge are very different things. Your goal here is to attain knowledge. Attained knowledge is power that is yours forever. But gaining this knowledge is your responsibility. It is something only you can do. Your success in this course depends not only on what we do for you, but more importantly, what you do for you.

We as an institution have made concerted efforts to create an environment in which you will succeed and learn. You are reading this inexpensive, specially designed book containing text, student workbook, laboratory manual, and reference tables written with you and your sixteen week, four credit physics requirement specifically in mind. You have a fine classroom/lab with overhead projection, air and power, and several lab stations with lab equipment for multiple lab groups. You have a qualified instructor. Free tutoring is available. It is your responsibility to embrace these many resources and effectively use them with a positive and constructive attitude.

All successful students do certain things routinely that leads to their academic achievement. Here are the academic expectations:

- Read! Reading is a basic thing all successful students do, and do regularly. Anyone with an education reads. Read and reread every word in this book. Your book contains the text, student exercises with answers, laboratory manual, glossary, index, reference tables, and more. Use it!
- Complete all the Student Exercises with understanding as we progress through the material. Much of the material, especially the problem solving, is skill based, and skills can only be acquired with repeated practice. Also, the subject matter is "sequential", meaning concepts build as we progress, and the concepts covered later cannot be understood unless the earlier ones are understood, so do what you have to do to keep up

on a regular basis. Expect to devote at least two hours of work outside of class/lab for every hour in class/lab.

- Do your best to keep "unifying principles" in mind while plowing through all the details of the individual energy systems. The "systems approach" adopted in this text takes advantage of the fact that different "energy systems" (such as mechanical, fluid, electrical, etc.) are analogous, meaning they are alike in important and useful ways. For example, electrical systems are often compared to fluid systems, with electrical "current flowing" through wires much like liquids flowing through pipes. It is your job to try to keep these unifying ideas in mind while we work through all the different units of measurement, letters used in formulae, and other differences and details in the individual energy systems. (More on the systems approach and unifying principles later.)

- Complete the laboratory activities with understanding. Labs are done in small groups. Be a good lab partner by getting involved and making a positive contribution. Lab activities can be fun and engaging. They also are a great opportunity to interact with your peers and your instructor. The lab activities not only reinforce concepts covered in the text, but additional subject matter is introduced in some of the labs. Read the labs completely before starting the activity.

- Take advantage of instructor lectures. Attending classroom lectures day after day can indeed be dull and boring at times, but once again, lectures are a powerful and useful resource if you take advantage of them. Your instructor will do his best to lay out the course content in an organized way verbally and with demonstrations, and many examples. Take notes, dating each session. Be an active listener. Lean forward. Ask questions. (Making jokes and entering discussions off-topic is welcomed as long as it is appropriate and doesn't manipulate too much class time. This type of thing is part of positive classroom engagement.)

- Our school has what is called the Academic Success Center (ASC) where free tutoring is available. The ASC staff will be more than happy to accommodate you. They will schedule you as best they can with a qualified tutor. Take advantage of this valuable, free resource needed or not.

- In my experience, my best students have been people who have made our school part of their lives. Show up for class regularly, rested, groomed, and prepared. You belong here. You are part of our family. Get to know your teachers and staff. Be part of what's going on. Embrace your college experience as one that will ultimately be a life-changing and thoroughly enjoyable journey.

A Broad Understanding

Skilled technicians are a valued part of any technological society. Technicians keep our lights on, our cars rolling, our furnaces heating, our computers running, our roofs from leaking, and much, much more. And with the accelerating pace at which technologies change and progress, technicians must not only adapt and keep up with increasingly complex advances, but must also combine various types of technologies into ever more sophisticated systems. To meet these challenges, today's technicians must have a broad understanding of technical principles spanning multiple energy systems. This course is designed to provide such a broad and useful foundation.

Physics

Physics is the study of the material (physical) world. Topics include force, velocity, energy, and power. There are many types of science, but physics is the foundation of them all. Much of studying physics is learning the standards that are used to measure and describe the physical world. So we'll learn science's language, its ways of writing and communicating, and its various standards of measurement. All this provides the foundation from which technicians can read new technical material with comprehension, communicate with one another accurately and succinctly, and generally stay current and successful in their ever evolving and overlapping fields of expertise.

Science, and its foundation in physics, is a way of knowing. Philosophically, there are many things that must be taken on faith in order to do science, unprovable ideas we must simply believe in. These are called the **philosophical presuppositions** of science, things we "presuppose" ahead of time. For example we must believe that there are patterns and rules that nature always follows. Sometimes we call these rules the "laws" of nature. We must also believe that we humans are capable of understanding these rules and that we can use them to our advantage. There is reason to have faith in these presuppositions because science routinely produces demonstrable results.

But science also has its limitations. Science is not a trump card over other ways of thinking and does not guide other aspects of our lives. One major **limitation of science** is that science cannot provide its own ethical guidelines; it cannot tell you the difference between right and wrong. "Good" science is science that works regardless of how it is used. Quite often science gets ahead of us, creating controversial ethical issues such as cloning and other forms of genetic manipulations, stem cell research, and abortion. In each case the science is "good" science because it works. But whether or not these sciences should be pursued and implemented is a question that cannot be answered in the scientific arena, but answered elsewhere.

Quantities

Throughout this course we will often use the term "quantity". One definition for this word is how much of something there is. For example, there might be eighteen students in the classroom, a quantity of students. But there is another definition for this word, a technical definition. In physics, a **quantity** is any characteristic or property of the physical world that can be **quantified**, or measured. Mass, speed, heat, and voltage are all quantities. Note that we have not ascribed any number to them, but expressed them only as quantifiable, measureable properties.

> **Examples of Quantities**
> distance
> area
> torque
> velocity
> energy
> temperature
> power, etc…

There are two types of quantities, scalar quantities and vector quantities. Scalars are simpler than vectors. A **scalar quantity** can be described by only how much, it's **magnitude**. Mass, fluid volume, and distance are examples of scalar quantities. A **vector quantity** has magnitude like a scalar but also direction, a "which way" component as well. An example of a vector is force, which not only has magnitude, measured maybe in pounds, but also a **direction** at which the force is acting. Both magnitude and direction are necessary to fully describe a vector quantity like force. Other vector quantities include velocity and momentum.

> **SCALAR QUANTITY: MAGNITUDE ONLY**
>
> **VECTOR QUANTITY: BOTH MAGNITUDE AND DIRECTION**

The direction of a vector can be described roughly as "up" and "down" or maybe "north" and "south", but vector direction is more precisely described as an angle, usually represented by Greek letter theta (θ), and measured in degrees, revolutions, or radians. Radian measure is an important ratio to understand because its use is necessary in most formulae in rotational systems. A review of radian measure is included in the review section.

Physics is the foundation of all the sciences largely because it is in physics where much of the language and standards of units (such as the gram being established to quantify mass) have been established. Much of this course is devoted to developing a technical vocabulary. For example the words "pressure" or "stress" would be used in a very different sense in psychology than in physics. Developing a technical vocabulary and learning technical standards and conventions is a major goal of this course.

Energy Systems and the Systems Approach

Physics has traditionally been taught by treating technical topics as separate disciples, with chapters typically organized as two sequential four credit courses covering mechanics, heat, sound, wave motion, electricity, magnetism, and optics, usually in this order. This is the standard College Physics I and College Physics II sequence.

This traditional format has at least two significant drawbacks. The first is that the important similarities of the various mechanical, fluid, electrical, and thermal systems is not directly addressed, lacking in any broad understanding of how all these disciplines are interrelated or **analogous** (alike). Secondly, time spent on formulae derivations and proofs is done at the expense of establishing a sense of how the various systems can interact and how they are used in the scientific, commercial, and industrial applications.

Here much of the same content of a traditional physics course is covered except that the material is organized and presented in a very different way, one more appropriate for students in technical and trades programs. We will use **unifying principles** in a **systems approach**.

The natural world can be categorized in many ways, but one way useful to technicians is organizing technologies into **energy systems**. Technologies from automobiles to photocopiers are best divided into energy subsystems. Energy systems are groups of quantities that work together. We will study five energy systems. Although each of these five energy systems is a separate discipline, note how they are also analogous (similar) to one another, each with its quantity that acts as the "mover" and also each with a quantity that is being moved:

Analogous Energy Systems

In translational mechanical systems forces move objects.

In rotational mechanical systems torque turns objects.

In fluid systems pressure moves fluid.

In electrical systems voltage moves charge.

In thermal systems temperature difference moves heat.

Utilizing the **systems approach**, each chapter addresses a unifying principle that is applied to each energy system one by one. A **unifying principle** is a concept, idea or relationship that can be applied to all of the energy systems. *The unifying principle is "unifying" in that it unifies the energy systems.* Each chapter is then divided into five sections, one for each energy system. For example, the final chapter on power covers the unifying principle of power. Like the other chapters, sections in this chapter are divided by energy system, chapter number, section number.

This is a **sequential course**. This means that subject matter is presented in an order that builds up to later ones in later chapters, so it is crucial students complete each chapter with comprehension. For example, as discussed above, there is one formula for power, a unifying equation that can be applied to all of the energy systems, but this concept could never be understood without the understanding of the unifying principles that lead up to it. Our final exam is on power.

Literals and Units

Quantities are represented as letters in formulae, a shorthand that makes formula writing and manipulation much easier. The **literal** is the shorthand letter used in the formulae. It is very important to learn what these short-hand letters mean, otherwise it becomes difficult to understand which quantities are represented in formulae and to be able to create and write solutions to physics problems. Usually the literal is the first letter of the word, for example "F" is used to represent force, "A" is used to represent area, and "a" is used to represent acceleration. Sometimes the literal is not the first letter of the word. Sometimes Greek letters are used, particularly in rotational systems, such as theta (θ) for an angle.

Note there is an important distinction between **upper case** verses **lower case letters**, so get used to writing "A" for area but "a" for acceleration, "F" for force but "f" for frequency, and so forth. Also, sometimes letters can be confused with numbers, so cursive or italics letters are used. For example, lower case "l" for lever arm can be confused with the number "1", so a different looking "ℓ" might be used. All these conventions are designed to avoid ambiguity an eliminate confusion.

Units are the agreed-upon standards of measurement. Some of these have been handed down for many years by our ancestors. There are two systems of units, the **English system** and the far superior **metric system**. The metric system is commonly called the **SI system**, short for "Systeme Internationale". The international scientific community met in France to set it up. Much like the quantity literals discussed above, there are shorthand letters that represent these units. For example, grams is "g", pounds is "lb", and meters is "m".

While on the subject of literals, it's worth mentioning there are another group of shorthand letters that represent power of ten. The student should be familiar with these for it is part of the math prerequisite. For example, kilo means $\times 10^3$ and is represented with "k", mega means $\times 10^6$ and is represented with "M". These prefixes are reviewed in review chapter below.

At first it can be confusing when trying to keep track of the "alphabet soup". This is one of the many challenges of our course of study. After a while it begins to make sense. For example, in d = 12.5mm, from left to right, "d" is for distance, "m" is for milli, and the second "m" is for meters. There are two lower case m's here, each meaning something different!

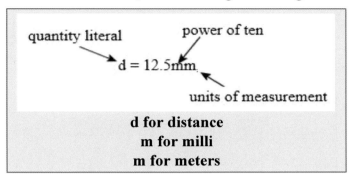

d for distance
m for milli
m for meters

Note that the meaning of the letter depends on its location in the statement; the quantity literal is at the left, then after the equal sign the units follow the number, but a power of ten might be just in front of the units. In order they always appear as the literal − equal sign − power of ten − units.

Another example is m = 13.6 mg. From left to right, quantity "m" for mass (another m!), the equals sign, power of ten "m" for milli, then units "g" for grams.

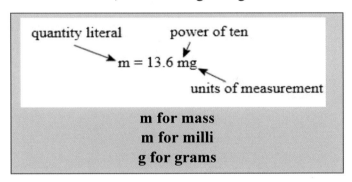

m for mass
m for milli
g for grams

Sometimes different values of the same type of quantity can appear in the same situation, say a bunch of different electronic resistors connected together, but each represented by the letter "R". To tell all the different R's apart, **subscripts** are used. A subscript is a smaller letter or number written to the bottom right of the literal. In the case of the resistors, each can be individually represented by R_1, R_2, R_3, and so on. Or two velocities, one initial and one final, could be written v_i and v_f.

The subscripts are a naming convention, not a numerical operation. On the other hand, **exponents** are numerical operations written in **superscript** position. For example, $2^3 = 8$.

Superscripts are often used for another purpose, a shorthand for units of measurement. These are not numerical operations. For example, V = 12.5 cubic feet can be written V = 12.5 ft^3. Note that the superscript three is not a numerical operation. It is "on" the ft units. The 12.5 is not cubed. ft^3 is short for cubic feet.

We'll work on these one bit at a time, introducing literals, units, formulae, and unifying principles as we go along chapter by chapter and section by section.

**Introduction
Key Terms & Phrases**

Physics
Philosophical Presupposition
Quantity
Quantify
Scalar
Vector
Unifying Principle
Energy System
Systems Approach
Analogous
Subscript
Superscript
Exponent
Power of Ten
Literal
Units of Measurement
English System
SI System
Sequential Course of Study

R   REVIEW OF PREREQUISITE STUDIES

A prerequisite course is a course that must be satisfactorily completed prior, or before, another course. These courses prepare you for the more difficult upper level courses required to complete your degree. Physics courses almost always require a math prerequisite because in physics, like most sciences, math is used as a tool. Skills such as number systems, computations, formulae, and power of ten are used routinely. Physics is not a math course. Physics is about other things. Math is a tool we use to do physics.

Math is only one type of prerequisite academic skill. Other skills include reading with comprehension, engaged listening, speaking articulately, time management, active participation, computer skills, taking personal responsibility for one's own work, and other proficiencies.

You would not be enrolled in this physics course without having satisfactorily completed a prerequisite math class. So let's review some of these prerequisites, particularly the math.

R1   Converting Units – Unit Factor Method

Units of measurement are very important and must accompany the answer to any calculation. It isn't just about crunching numbers. These numbers mean something. They quantify concrete things in the material world. The number represents how many of a particular standard of measure there is; feet, psi, etc. For example, the number 12 does not equal the number 1, but 12 inches does indeed equal 1 foot. As a result, a number without accompanying units of measurement is meaningless in physics. The only exception are the ratios, which, by definition, have no units.

There are many instances in which units of measurement of the same quantity must be converted to other units of measurement. For example, a simple conversion might be changing a given number of miles to an equivalent number of feet. Or maybe the mileage is converted to units in the metric system, say kilometers.

The reasons units must be converted depends on the situation: contracts, regulations, standards, specifications. But more importantly, while making calculations using formulae, and generally solving problems in science, requires that the units of measurement be **consistent**. Keeping the units consistent means that they must all combine together in a legitimate manner, with some units maybe cancelling out, in order to get units in the answer that makes sense. For example, let's calculate how much carpet is needed to completely cover a room that is 20 feet by 10 feet. The area of the room is calculated using the formula $A = LW = (20\text{ft})(10\text{ft}) = 200\text{ft}^2$. The point is that feet must be multiplied to feet to get square feet. The units in A =LW must be the same, or *consistent*. When adding, feet must be added to feet to get feet, once again consistent.

So converting units is an important skill that is often utilized when solving science problems. Here we will learn a method for converting units that can be very useful, especially when the units are unfamiliar. This method is called the **unit factor method**.

The unit factor method makes use of the fact that any value does not change when multiplied by the number one. A simple example will help illustrate this method. (Now there's no doubt the reader can readily convert inches to feet without the need of a step-by-step method, but the point here is to understand the process for when the conversions are not so obvious or familiar.)

Let's take a very simple example and convert 1.5 feet to an equivalent number of inches. What must be known is how many inches are in a foot or how many feet there are in an inch. Either way will work. If this relationship is not immediately known then it is looked up in a conversion table: one foot equals twelve inches. The question now is whether to divide the given 1.5 feet by 12 or to multiply by 12. Most people understand that since the inch is smaller than the foot, more of them are needed to be equal, so the 1.5 must be multiplied by twelve to get a bigger number. This works great when the units are familiar, but things might get a bit more confusing when the units aren't so easy, obvious, and require multiple conversions. This is where the unit factor method comes in handy.

A factor is a number that is multiplied. The **unit factor** is a fraction that is used to covert units. *When the numerator and the denominator of a fraction are equal, the value of the fraction is numerically equal to one.* The trick with our unit fraction is that while the numerator and denominator are equal, they are not of the same units. We build this fraction a certain way to convert, ultimately always multiplying by one.

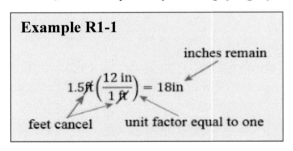

Example R1-1

inches remain

$$1.5\,\text{ft}\left(\frac{12\,\text{in}}{1\,\text{ft}}\right) = 18\text{in}$$

feet cancel   unit factor equal to one

Given 12 in = 1 foot, the unit factor is created by substituting one of these values in the numerator and the other in the denominator. Either way, our unit fraction is equal to one. But they must be *arranged such that the initial units must cancel*, here feet, so feet must be in the denominator. Inches must be in the numerator. As a result there is no doubt that in order to covert from feet to inches, the given value in feet must be multiplied by the twelve inches and divided by the one foot. Yes, it is true that dividing by one does nothing numerically, but it is important to understand that by dividing by one foot, feet cancels, with inches remaining above in the numerator. Also, there are situations where there might not be a value of one in the unit fraction.

Concise conversion tables will contain many conversion factors in a small space by listing the unit equivalents beginning with a value of one, then expanding to many other units of measurement for the same quantity. Any two in the list can be paired up into a unit factor. For example,

$$1\ \text{h.p.} = 746\ \text{W} = 550\ \frac{\text{ft--lb}}{\text{sec}} = 2545\ \text{Btu/hr} = 178\ \text{cal/sec}$$

The beauty of the unit factor method is that you don't even have to know what type of quantity is involved (here power) or what any of the units mean; you're just converting using the given equivalents and applying the unit factor method. For example, let's convert 1200W to Btu/hr. Example R1-2 shows how the conversion factor is constructed using the above list of equivalents. The unit factor shows we must multiply by 2545 and divide by 746 in order to cancel "W" and get the desired Btu/hr.

Even with these unfamiliar units, we are able to correctly convert the them by creating a unit factor; note carefully how the units that must cancel are in the denominator and the units that must remain, the desired units, are in the numerator. We then know we must multiply the 1200 by 2545 and divide by 746. Remember the unit factor is equal to one because, even though the units are different, its numerator equals its denominator.

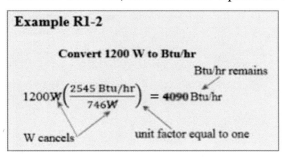

Example R1-2

Convert 1200 W to Btu/hr

$$1200 \text{W}\left(\frac{2545 \text{ Btu/hr}}{746 \text{W}}\right) = 4090 \text{ Btu/hr}$$

Btu/hr remains

W cancels   unit factor equal to one

Fundamental and Derived Units

There are three **fundamental quantities** which have **fundamental units**; mass, length, and time. All other quantities, along with their respective standards of measurement, are combinations of the three fundamental quantities, or **derived** from the three basic quantities.

Fundamental Quantity	English Units	SI units
mass	slug	gram
length	foot	meter
time	second	

For example, isn't area the mathematical product of two lengths? Isn't speed a division of length and time? All quantities, and their resulting units, are combinations of these three fundamental quantities. The units that result from mathematically combining the fundamental units, like ft^2 for area and ft/sec for speed, are called **derived units**.

Example R1-3

Convert 50 mi/hr to ft/sec using only fundamental conversion factors.

Solution: $50 \text{mi}\left(\frac{5280 \text{ft}}{1 \text{mi}}\right) = 264{,}000 \text{ ft}$

$1 \text{hr}\left(\frac{3600 \text{sec}}{1 \text{hr}}\right) = 3600 \text{ sec}$

$\frac{50 \text{ mi}}{1 \text{ hr}} = \frac{264{,}000 \text{ft}}{3600 \text{sec}} = 73.3 \text{ ft/sec}$

Converting derived quantities with derived units can be done by using the fundamental quantities. For example, converting miles per hour to feet per second can be done by converting the two fundamental quantities of length and time separately, then combining them after; we can convert miles to feet and hours to seconds then divide the two values. As a result, long lists of conversion factors become unnecessary because most conversions can be done by knowing only a few fundamental conversion factors.

Using the table of equivalents in the reference table provided in the back of this book, page 354, convert the given units. Write out the unit fraction and its cancelled units.

1. Convert 32.0 inches (in) to meters (m).

2. Convert 10.0 miles (mi) to kilometers (km).

3. Convert 562 cubic inches (in^3) to gallons.

4. Convert 1000 Btu/hr to watts (W).

5. Convert 288 inches to feet (ft).

6. Convert 288 square inches to square feet(ft^2).

7. Convert 288 cubic inches to cubic feet(ft^3).

8. Convert 10.0 psi to pounds per square foot $\left(\dfrac{lb}{ft^2}\right)$.

9. Convert 45.5 centimeters (cm) to meters.

10. Convert 15.2 milliamps (mA) to amps.

Answers to Unit Conversions
1. 0.813 m
2. 16.1 km
3. 2.43 gal
4. 293 W
5. 24.0 ft
6. 2.00 ft^2
7. 0.167 ft^3
8. 1440 $\dfrac{lb}{ft^2}$
9. 0.455 m
10. 0.0152 A
11. 71,500 ft-lb/sec
12. 6.23 hp
13. 0.1245 m^2
14. 77.1 kg
15. 1.20 kΩ
16. 88.0 ft/s
17. 8.90 N

11. Convert 130 horsepower (hp) to foot-pounds per second (ft-lb/sec).

12. Convert 4650 watts to horsepower (hp).

13. Convert 1245 square cm (cm^2) to square meters (m^2).

14. Convert 170 pounds (lb) to kilograms (kg).

15. Convert 1200 Ω (Ohms) to kΩ.

16. Convert 60.0 mi/hr to ft/s.

17. Convert 2.00 pounds to Newtons (N).

Formulae are mathematical statements of equality that express relationships between quantities. Without them, expressing these relationships would be tedious, imprecise, and cumbersome. In the Introduction above, we learned about literals, letters that are used to express various quantities without having to write the whole word. For example, "force" is represented with a capital "F". Relationships between two or more quantities are expressed in mathematical form using these literals.

A commonly known formula is $V = LWD$, the formula for calculating the volume of a rectangular solid (the shape of a brick), where the symbol V represents the volume, L represents the length, W represent the width, and D represents the depth. This formula states that the length, width, and depth must be multiplied together to calculate the volume in a tight, efficient package without having to write out whole sentences. We can also apply algebra to perform the indicated mathematical operations.

> The plural for the word formula can be written either formulas or formulae, the latter more accepted in technical literature.

When the symbols are butted up against each other with nothing in between them, they are to be multiplied together. The division symbol $\div$ is avoided. In formulae, division is written in fraction form, such as in $P=F/A$ or $P = \dfrac{F}{A}$.

In $V = LWD$, V is the **isolated literal**. This means that this is the letter by itself to one side. Values for L,W, and D can be **substituted** (plugged in) and multiplied to calculate the volume. But what if the volume (V) is given and one of the other quantities is to be found, say if the length (L) was to be calculated. This formula would then have to be manipulated so that L is isolated instead of V. **Formula manipulation** is the algebra we apply to isolate the quantity that is to be found.

In algebra we learned how to solve for an unknown in an equation by applying the property of equality. By performing the same operation to both sides of an equation, the equation remains true, equal, thereby isolating the unknown to one side to determine its value. In the example shown, the same mathematical operations are performed to both sides to isolate x. We learned that the **inverse** appears on the other side of the equation: plus three on the left becomes negative three on the right and while the two is multiplied to the unknown on the left, it is divided into the six on the right. Solving equations is a game of inverses, undoing what is happening to the unknown to get it by itself. Manipulating formulae is done the same way except there aren't necessarily any numbers involved, all literals representing a relationship between quantities.

> **Solve for x**
>
> $$2x + 3 = 9$$
> $$2x = 9 - 3$$
> $$2x = 6$$
> $$x = \frac{6}{2}$$
> $$x = 3$$

To isolate L in $V = LWD$, we must divide both sides by W and also D for them to cancel from the right side of the equation thereby isolating L.

The result is $\dfrac{V}{WD} = L$, but usually the isolated literal is written on the left side. The equation remains equal when we flip the sides, $L = \dfrac{V}{WD}$.

Since nearly all of the formulae used in this course are quantities that are multiplied and divided, a useful little trick might be worth mentioning. Remember that letters can be flipped to the other side of the equal symbol as an inverse. Simple formulae with only multiplication and division operations can be manipulated easily and quickly using this principle. The method is to simply remove letters from one side of the equation and writing them on the other side as an inverse. The following examples illustrate this "switch-a-roo". It is not necessary here to know what the letters mean, just to manipulate the formula to isolate a given letter.

Example R2-1 Isolate F in W = Fd

$$\frac{W}{d} = \frac{F\cancel{d}}{\cancel{d}}, \qquad \frac{W}{d} = F, \qquad F = \frac{W}{d}$$

We first divide both sides by d. d *is in the numerator multiplied to* F, *so is moved to the other side in the denominator under the* W.

Example R2-2 Isolate E in P = EI

$$E = \frac{P}{I}$$

I *is in the numerator multiplied to* E *so is moved to the other side in the denominator under the* P.

Example R2-3 Isolate F in $P = \dfrac{F}{A}$

$$F = PA$$

A *is in the denominator under the* F *so is moved to the other side in the numerator multiplied to the* P.

Example R2-4 Isolate h in P = ρgh

$$h = \frac{P}{\rho g}$$

ρ *and* g *are in the numerator multiplied to the* h *so they are both moved to the other side in the denominator under the* P.

Instructions: It is not necessary for the student to know what the formulae or individual letters represent, only to be able to understand the mathematical relationships between the letters and to algebraically manipulate the formula for any required letter. "Manipulating a formula" means isolating the unknown letter in the formula. These exercises are designed to practice these algebra skills. The formulae shown are those the student will encounter in this course.

1. Isolate m in $w = mg$.

2. Isolate θ in $W = \tau\theta$

3. Isolate h in $p = \rho h$

4. Isolate F in $p = {}^F\!/_A$

5. Isolate A in $p = {}^F\!/_A$

6. Isolate V in $\rho = {}^m\!/_V$

7. Isolate d in $P = \dfrac{Fd}{t}$

8. Isolate t in $P = \dfrac{Fd}{t}$

9. Isolate ΔT in $Q = mc\,\Delta T$

10. Isolate p_g in $p_T = p_{atm} + p_g$

11. Isolate v_f in $v_{avg} = \dfrac{v_f + v_i}{2}$

12. Isolate W_{out} in $\eta = \dfrac{W_{out}}{W_{in}}$

13. Isolate W_{in} in $\eta = \dfrac{W_{out}}{W_{in}}$

14. Isolate I in $E = IR$

15. Isolate r in $A = \pi r^2$

16. Isolate V in $P = \dfrac{V^2}{R}$

17. Isolate $\Delta\omega$ in $\alpha = \dfrac{\Delta\omega}{t}$

18. Isolate t in $a = \dfrac{v_f - v_i}{t}$

19. Isolate d in $a = \dfrac{2d}{t^2}$

20. Isolate b in $c^2 = a^2 + b^2$

Answers to Formula Manipulations

1. $m = \dfrac{w}{g}$ **2.** $\theta = \dfrac{W}{\tau}$ **3.** $h = \dfrac{p}{\rho}$ **4.** $F = pA$ **5.** $A = {}^F\!/_p$ **6.** $V = {}^m\!/_\rho$ **7.** $d = \dfrac{Pt}{F}$

8. $t = \dfrac{Fd}{P}$ **9.** $\Delta T = \dfrac{Q}{mc}$ **10.** $p_g = p_T - p_{atm}$ **11.** $v_f = 2v_{avg} - v_i$ **12.** $W_{out} = \eta W_{in}$

13. $W_{in} = \dfrac{W_{out}}{\eta}$ **14.** $I = \dfrac{E}{R}$ **15.** $r = \sqrt{\dfrac{A}{\pi}}$ **16.** $V = \sqrt{PR}$ **17.** $\Delta\omega = \alpha t$

18. $t = \dfrac{v_f - v_i}{a}$ **19.** $d = \dfrac{at^2}{2}$ **20.** $b = \sqrt{c^2 - a^2}$

Power of Ten

There are many occasions, especially in science-related topics, where we may come across either very large or very small numbers. For example,

- Residents of New York City use a total of 180,000,000,000 gallons of water per year.

- The mass of the Earth is approximately 5,980,000,000,000,000,000,000,000 kg.

Writing numbers in decimal form like this, with all those zeroes, is messy, tedious, and prone to errors. But there's a better, more efficient way to write such numbers that make them easier to work with. This other form is **power of ten**. "Power" is a term often used for an exponent. For example, in 2^4, two is the called the base and four is the exponent. The two is multiplied to itself four times. (Careful, this is not the same as multiplying two by four.) 2^4 might be pronounced "two to the fourth power". When the base is ten, and ten is raised to a power, the decimal point simply moves back and forth. As a result, this is an excellent way to handle those pesky zeroes.

Before we discuss this method, let us review powers of ten using integer exponents at right.

Whenever we multiply or divide by ten, the decimal point moves one decimal place value. The exponent indicates how many places the decimal point moves.

Using power of ten, it becomes easier to write, say, and make computations with very small or very large numbers. For example, using power of ten:

$$10^3 = 10 \cdot 10 \cdot 10 = 1000$$
$$10^2 = 10 \cdot 10 = 100$$
$$10^1 = 10$$
$$10^0 = 1$$
$$10^{-1} = \frac{1}{10} = 0.1$$
$$10^{-2} = \frac{1}{10^2} = \frac{1}{100} = 0.01$$
$$10^{-3} = \frac{1}{10^3} = \frac{1}{1000} = 0.001$$

- fifty-two million liters of water spill over Niagara Falls per day.

$$52,000,000 = 52 \times 10^6 \text{L/day}$$

- a paper manufacturer produces paper consistently 0.000096 meters thick
$$0.000096 = 96 \times 10^{-6} \text{m}$$

Scientific Notation

Scientific notation is a particular style of writing large or small number in power of ten. To be in scientific notation, the decimal point must be in a particular place; only one digit can be to the left of the decimal point. Using the examples above,

- $52{,}000{,}000 = 5.2 \: x \: 10^7$

- $0.000096 = 9.6 \: x \: 10^{-5}$

One advantage to using power of ten notation is the ease with which calculations can be made using the rules for exponents. For example, not how the zeroes cancel out by inspection (without the use of a calculator) when the huge values are multiplied or divided:

- $$\frac{3.6 \text{ x } 10^{15}}{1.2 \text{ x } 10^{11}} = 3.0 \text{ x } 10^4$$

- $(2.0 \: x \: 10^3)(6.0 \: x \: 10^5) = 1.2 \: x \: 10^9$

It's also very easy to handle these very large and very small numbers when using a scientific calculator. On the TI-30, the EE key is used to enter power of ten. It is the second function of that key, so "2nd" must be pressed first. Other calculators like the Casio label a key EXP for power of ten. Practice using your own calculator so that you can confidently enter and make computations with large and small numbers using you power of ten key.

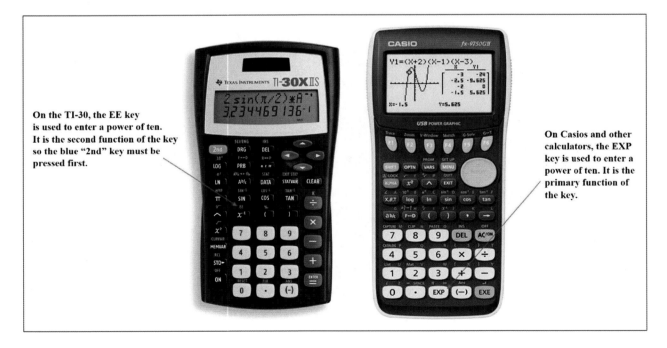

On the TI-30, the EE key is used to enter a power of ten. It is the second function of the key so the blue "2nd" key must be pressed first.

On Casios and other calculators, the EXP key is used to enter a power of ten. It is the primary function of the key.

Integer Exponents

Recall the definition of exponential notation, where the exponent represents the number of times the "**base**" is multiplied to itself:

$$x^n = x \bullet x \bullet x....$$

As a result when a common base is multiplied or divided we can add the exponents when the common base is multiplied and subtract the exponents when the common base is divided:

$$x^n \bullet x^m = x^{n+m} \quad \text{and} \quad \frac{x^n}{x^m} = x^{n-m}$$

To use power of ten efficiently when performing operations with large and small numbers, the above two **rules of exponents** make handling the zero place holders a simple, error-free process. These two rules are the **product rule**, when common bases with exponents are multiplied, and the **quotient rule**, where common bases with exponents are divided. When using scientific notation, the "common base" is always number ten.

Example R3-1
$$10^3 \bullet 10^5 = 10^{3+5} = 10^8$$

Example R3-2
$$10^8 \bullet 10^{-2} = 10^6$$

Example R3-3
$$(3 \times 10^5)(2 \times 10^3) = 6 \times 10^8$$

Example R3-4
$$\frac{10^9}{10^6} = 10^3$$

Example R3-5
$$\frac{10^{18}}{10^{-10}} = 10^{28}$$

Example R3-6
$$\frac{8 \times 10^6}{4 \times 10^2} = 2 \times 10^4$$

- When applying these rules to powers of ten, the base is always ten. When the powers of ten are multiplied, the exponents are added. This sum is the total number of times the ten is multiplied to itself.

- When the powers of ten are divided, the exponents are subtracted, with some of the zero place holders cancelling.

Metric Prefixes

The English system of units is a cumbersome bunch of numbers. There are 12 inches in a foot, 3 feet in a yard, and 5280 feet in a mile, to name a few. When the metric system was devised, units were established to make conversions easier. They used power of ten.

To make writing and saying powers of ten easier, the powers of ten were given a name and an abbreviation. These are called **metric prefixes** because the abbreviation of the power of ten precedes the units of measurement. There are five of these you should know by heart.

Make a distinction between upper and lower case letters, with capital "M" for mega and lower case "m" for milli. The μ symbol for micro is the Greek letter mu.

Metric Prefixes	
micro (μ)	x 10^{-6}
milli (m)	x 10^{-3}
centi (c)	x 10^{-2}
kilo (k)	x 10^{3}
Mega (M)	x 10^{6}

As already mentioned, the metric power of ten prefix is written in front of the units of measurement.

For example,

$$0.00000254 \text{ A} = 2.54 \times 10^{-6} \text{ A} = 2.54 \mu A$$
$$0.075 \text{ m} = 75 \times 10^{-3} \text{ m} = 75mm$$
$$0.067m = 6.7 \times 10^{-2} \text{ m} = 6.7cm$$
$$2,500,000Pa = 2500 \times 10^{3} \text{ Pa} = 2500kPa$$
$$750,000,000 \text{ V} = 750 \times 10^{6} \text{ V} = 750 \text{ MV}$$

At this time it is not necessary to understand exactly what the units mean, only that they are units of measurement are preceded by a metric power of ten prefix in order to more easily say, write, and compute with numbers that have many zero placeholders.

Perform the indicated operations using the rules for exponents. Leave answer in exponent form.

1. $10^{-4} \cdot 10^7 =$

2. $10^6 \cdot 10^5 \cdot 10^{-3} =$

3. $\dfrac{10^{-4} \, (10^2)}{10^{-3}} =$

4. $\dfrac{10^4 \cdot 10^{-5}}{10^{-7} \cdot 10^{-2}} =$

Rewrite the numbers in scientific notation.

5. $8245 \, N =$

6. $93{,}000{,}000$ miles $=$

7. $0.0054 =$

8. $0.000\ 000\ 072 =$

Rewrite the numbers in standard form (without powers of ten).

9. $1.62 \times 10^3 =$

10. $4.4 \times 10^0 =$

11. $8.76 \times 10^{-5} =$

12. $2.1 \times 10^6 =$

Perform the indicated operations - by inspection - using the product rule and quotient rule. Leave the answer in scientific notation.

13. $(6.0 \times 10^3)(2.0 \times 10^4) =$

14. $(1.5 \times 10^6)(2.0 \times 10^{-2}) =$

15. $\dfrac{(5 \times 10^{-4})(2 \times 10^2)}{2 \times 10^{-3}} =$

16. $\dfrac{1.2 \times 10^{120}}{2 \times 10^{118}} =$

Perform the indicated operations using the scientific notation function on your scientific calculator.

17. $\dfrac{(6.21 \times 10^{-4})(3.55 \times 10^2)}{1.2 \times 10^{-3}} =$

18. $(8.21 \times 10^8)(2.24 \times 10^{-2}) =$

Answers to R3 Exercises

1. 10^3 **2.** 10^8 **3.** 10 **4.** 10^8 **5.** 8.245×10^3 N **6.** 9.3×10^7 mi **7.** 5.4×10^{-3}
8. 7.2×10^{-8} **9.** 1620 **10.** 4.4 **11.** 0.0000876 **12.** $2{,}100{,}000$ **13.** 1.2×10^8
14. 3.0×10^4 **15.** 5×10^1 **16.** 6×10^1 **17.** 1.8×10^2 **18.** 1.84×10^7

It is often said that machine parts must always be manufactured to high precision and that parts must always be made "perfectly". But is this true? No.

Nothing can be made "perfectly" because there are no perfect tools or perfect instruments to measure with. Parts are made only as precisely and accurately as they need to be. Some parts need to be made very precisely, some not. Parts are made to the precision and accuracy specified by the customer. It need not be made any better, otherwise the part is too expensive to make.

Measurements with tools such as pressure gages, calipers, and thermometers are never exact; they have limited precision and accuracy.

- All measurements taken with instruments (voltmeters, rulers, gages, etc.) are approximate. No measurement is exact.
- The final result of a calculation made with approximate numbers cannot be written with any more precision or accuracy than is present in the original measured values used.

Exact verses Approximate Numbers

Nearly all numbers used in technical work are **approximate numbers**, having been determined by a *measurement* from an instrument of some sort. The precision and/or accuracy of approximate numbers are limited to what the instruments can provide. None of these measured numbers are exact.

Numbers determined by a *counting process* are **exact numbers**. These are usually whole numbers.

Accuracy (Significant Digits)

The **accuracy** of an approximate number is the number of **significant digits** it has, that is, the total number of digits an instrument can provide regardless of place value. An approximate number may have to include some zeroes to properly locate the decimal point, zeroes actually measured by the instrument like it would any other digit. *Except for this zero placeholders, all other digits are significant digits.*

Accuracy is the number of digits of a measurement, the number of "significant" digits.

Zeroes in Measurements

A digital multi-meter display is the best way to begin understanding the accuracy of a measurement from an instrument. Meter display **#1** shown, is 686.8 VAC. What is important to understand is that this tool can provide no more than four significant digits; it's the maximum accuracy this tool can provide. All four digits are used here, so this measurement has an accuracy of four significant digits.

1.
Four Significant Digits

2.
Four Significant Digits

It's easy to count the significant digits when there are no zeroes. But when zeroes are in the measurement, some of the zeroes could be significant and some not. The difference is that *a zero that was measured by the tool is significant* while *zero place holders are not significant.*

There is one situation where a zero digit is always significant (measured). This is any zero, or series of zeroes, that is sandwiched between non-zero digits. In meter display **#2** shown, the meter reading is 606.8 VAC. The zero is between non-zero digits, so the accuracy of this measurement is four significant digits.

Another situation when the zeroes are always significant is when a zero, or series of zeroes, is positioned just to the right of the decimal point. These are the tenths, hundredths, etc. positions. These are not sandwiched between non-zero digits, but they were indeed measured by the tool and are significant. It just happens to be a measurement of zero instead of one through nine. Meter display **#3** shows such a reading. Both of these readings is four significant digits. Note that the placement of the decimal point is largely irrelevant; four digits are provided by the meter in both cases. *The zeroes to the right of the decimal point do not affect the pure numerical value of the number, but increase the accuracy of the measurement.* These measured digits in these place value positions simply just turned out to be zero measurements and are significant.

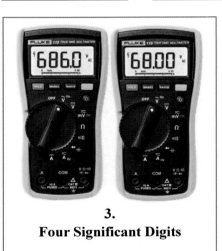

3.
Four Significant Digits

There is what is called the ambiguous case when it is impossible to know if a zero is significant or not unless one knows something about the tool or is otherwise told outright. Take for example a measurement such as 65,000 feet. Were the zeroes measured? Or did the technician have to write those zeroes in to make it sixty-five thousand, not sixty-five? Was the meter in kilo range? In order to communicate the accuracy of this measurement, the technician writing the number must indicate somehow what the accuracy is, 4. Four Significant Digits, that is, which of the zero digits is significant and which are not. This can be done simply by saying so, for example, "65,000 feet measured to an accuracy of ten feet". In other words, the right most zero in the ones position is a place holder. Another way of indicating accuracy is to write a line

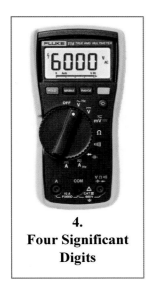

4.
Four Significant Digits

5.
Three Significant Digits

above the right-most significant digit. (This also can mean that there is a repeating decimal.) The technician can then write the measurement as 65,0$\overline{0}$0 feet. Note that the left-most zero need not have a line above; it's sandwiched between significant digits. 65,0$\overline{0}$0 feet has four significant digits.

On the other hand when the meter displays all these zeroes, there is no ambiguity because they've all been measured. Remember that in the ambiguous case, we're reading a number that was recorded, not the tool itself. There is none of this confusion when reading directly from the tool. For example in meter reading **#4** shown, the measurement of 6000 has accuracy of four significant digits.

The main reason scientific notation, the particular style of writing powers of ten, is so commonly used and especially useful, is that zero placeholder ambiguity in recorded measurements is eliminated, also eliminating the need of the lined zeroes. For example, in 2.30 x 10^3, the zero need not be written to properly position the decimal point. The place holders are handled by the power of ten. As a result, it is clear the technician measured that zero and the measurement contains three significant digits.

All of the examples given so far of the measurements with this digital meter have had accuracy of four significant digits. Certainly this particular meter can provide no more than four, but are there situations when it can be less? Meter reading **#5** is an example of such a reading. In most meters the left-most digits are not displayed when unused. Here the left-most digit is not being utilized, leaving only three digits available in the display. The accuracy of this measurement is three significant digits.

Zero place holders are not significant.

Precision

The **precision** of an approximate number is the smallest place value the instrument can provide. This place value is usually the *right-most significant digit*, the smallest place value the instrument can measure.

> **Precision is the smallest place value of a measurement.**

Determining the Accuracy and Precision of an Approximate Number

Imagine using various tools of some kind, both digital and analogue, which provide measurements of various precision and accuracy. Each measurement provides a particular precision and accuracy depending on the tool used and the skill of the user. Study the examples below so that given a particular measurement, you can **state its precision and accuracy**.

Sample Questions: Determine the Precision and Accuracy of a Measurement

Example R4.1 28.2 Accuracy: three significant digits, Precision: tenths

Example R4.2 280.2 Accuracy: four significant digits, Precision: tenths

Example R4.3 2802 Accuracy: four significant digits, Precision: ones

Example R4.4 2802.0 Accuracy: five significant digits, Precision: tenths (The zero is not a place holder, but was measured.)

Example R4.5 1600 Accuracy: two significant digits, Precision: hundreds (The zeroes are place holders unless otherwise told or if more is known about the measuring instrument.)

Example R4.6 1.6000 Accuracy: five significant digits, Precision: ten-thousandths (All those zeroes are not places holders, but were measured.)

Example R4.7 0.056 Accuracy: two significant digits, Precision: thousandths (The zero to the left of the decimal point is written to make the decimal point more visible. The zero to the right of the decimal point is a place holder.

Example R4.8 $12\bar{0}0$ Accuracy: three (because of the line over the zero), Precision: tens

Example R4.9 $1,20\bar{0},000$ Accuracy: four, Precision: thousands

Example R4.10 5.010×10^4 Accuracy: four, Precision: hundreds (or $501\bar{0}0$)

Direct verses Indirect Measurements

Direct measurements come directly from the instrument. These are always a comparison to standards of units of measurement.

Indirect measurements come from combining direct measurements in mathematical operations, usually with formulae.

Examples of Direct vs Indirect Measurements

Example R4.11   The floor area of a room can be determined indirectly with two direct measurements, the length and the width. The two direct measurements are multiplied using $A = L \times W$ to determine the area indirectly.

Example R4.12   Electric power can be determined indirectly by multiplying two direct measurements, the voltage and the current ($P = V \times I$). Power is indirectly determined.

Operations with Approximate Numbers (Rounding)

The right-most significant digit of a number is not exact. It has usually been determined by estimating or *rounding. The precision and accuracy of the result of a calculation is limited to the precision and accuracy of the original approximate numbers used.*

> When *adding and subtracting*, the result is limited to the *least precise* number used. The precision of any sum or difference is the precision of the least precise number used.

Examples of Rounding When Adding and Subtracting - Precision

Example R4.13   1.23 + 2.2 = 3.4 The least precise is 2.2 with precision of tenths. The sum is rounded to tenths from 3.43 to 3.4.

Example R4.14   1.26 + 2.2 = 3.5 The least precise is once again 2.2 with precision of tenths, but this time the right-most digit must be incremented, rounded from 3.46 to 3.5.

Example R4.15   1230 + 500 = 1700 The least precise is the 500 with precision of hundreds so 1730 is rounded to hundreds or 1700. (It is possible for these "ambiguous" zeros to be significant, but if so, should be stated somehow. For example, in Ex 13 below the precision of an altimeter is known to be precise to the nearest meter.)

Example R4.16   +   =   Here we are told the altimeter is precise to the nearest meter. These ambiguous zeroes are often noted with a line above the zero. This means that the zeroes were actually measured by the instrument and are not simply place holders. As a result, this answer is precise to the nearest one meter. We indicate this precision with a line above the zero.

Example R4.17   12,340 + 0.000002 = 12,340 The extremely precise number is "negligible" compared to the much larger number and does not change it.

Example R4.18   0.0342 − 0.006 = 0.028 The least precise is 0.006 with thousandths so the difference 0.0282 is rounded to 0.028.

When *multiplying and dividing*, the result is limited to the *least accurate* number used. The accuracy of any product or quotient is the accuracy of the least accurate number used.

Examples of Rounding When Multiplying and Dividing - Accuracy

Example R4.19 (6.00)(2.00) = 12.0 The two factors have accuracy of three significant digits so the product has accuracy of three significant digits.

Example R4.20 (6.00)(2.0) = 12 The least accurate number is 2.0 with two significant digits so the product has two significant digits.

Example R4.21 (6.00)(2) = 10 The least accurate number is 2 with one significant digit so the product 12 is rounded to one significant digit, 10, with the zero a place holder.

Example R4.22 (0.000234)(1.6) = 0.00037 The least accurate number is 1.6 with two significant digits so the answer is rounded from 0.0003744 to 0.00037. (Those fours were not measured. We simply do not know those values.)

Example R4.23   =   Here the highly accurate value for   is rounded to match the accuracy of the approximate number, 3.1. The approximate number 2.7 has two significant digits, so the answer can only have two significant digits.

Example R4.24 $(1\bar{0}0)(1.203) = 12$ There are two significant digits in $1\bar{0}0$, so the answer has two significant digits

Our Day-to-Day Rule-of-Thumb for Rounding

It's too much of a hassle to correctly round every mathematical operation we perform in every problem that's solved throughout our sixteen week course. So once students understand precision and accuracy, we simplify day-to-day calculations by always assuming, maintaining, and rounding to *three significant digits*. After moving on from the specifics of precision and accuracy, students will leave answers to three significant digits, and certainly no more than three significant digits. This rule-of-thumb does not apply during laboratory activities.

"Three Sigs" Rule-of-Thumb

Example R4.25 The pressure in a piston is 100 psi. Its cross-sectional area is 1.5 square inches. Determine the force produced at its shaft.

Solution:
$$F = PA = (100\,psi)(1.5\,in^2) = 150\,lb$$

The purpose of the above problem is to solve a piston pressure problem *theoretically*, not experimentally. Note how three significant digits are assumed regardless of the precision and accuracy of the given values. And no line is drawn over the zero in the answer 150. Are we breaking the rules for rounding? Not quite. These values were never actually measured with a tool, just made up for academic practice. It can be a real hassle concerning ourselves with rounding correctly all the time, so other than when doing laboratory activities where values are actually measured, this "three sigs" rule-of-thumb is used for the sake of convenience.

Percent Error

There are many lab activities where the experimental results of the activity are compared to what was expected or known. This is usually expressed as percent error. **Percent error** is the ratio of the difference between the experimental lab results and what was expected divided by what was expected, calculated by the formula shown. Note how the numerator is an absolute value difference with no concern of sign, just a difference. Here "error" does not mean a mistake was made.

$$\% \text{ error} = \frac{|\text{expected} - \text{experimental}|}{\text{expected}}$$

Example R4.26 Students have performed a laboratory activity where acceleration due do gravity is experimentally determined. The students' experimentally came up with $10.9 \, m/_{s^2}$. Acceleration due to gravity is known to be $9.81 \, m/_{s^2}$. Determine the percent error.

$$\% \, error = \frac{|9.81 - 10.9|}{9.81} = 0.111 = 11.1\%$$

In other words, the students' results were 11.1% off from what was expected (which isn't bad at all).

1. What is the precision and accuracy of the meter reading 1.470 meters?

2. What is the precision and accuracy of the measurement 0.0102 volts?

3. State whether the given number are approximate or exact.

    a.   A car with 8 cylinders goes 75 mph.

    b.   Half of 7200 seconds.

    c.   A calculator has 34 keys, and its battery lasted for 34 hours.

    d.   Forty-five students drove 150 miles.

4. Which of each pair is more accurate?

    a.   30.1 or 30.0

    b.   0.042 or 3450

    c.   7000 or 0.000023

    d.   128.0 or 128

5. Which of each pair is more precise?

    a.   16.02 or 0.03?

    b.   25 or 20?

    c.   25 or $2\overline{0}$?

    d.   0.020 or 0.02?

6. Round the numbers to three significant digits.

    a.   150,340

    b.   0.02356198

    c.   0.0449

    d.   1.00000000

    e.   0.6999

7. Perform the indicated operations, rounding appropriately.

    a. $3.8 + 0.154 + 47.26 =$

    b. $12.78 + 1.0495 - 1.633 =$

    c. $3.64(17.06) =$

    d. $(0.520)^2 =$

    e. $2\pi(18.08)^2 =$

8. The multi-meter used to measure electric current (I) in a circuit displays a value of 0.038 amperes. The same meter is used to measure 1215 Ohms resistance. Using the formula E = IR (Ohm's Law), calculate the voltage (E), rounding appropriately.

9. Explain in your own words why precision and accuracy can be extremely important when dealing with measured values, especially in manufacturing situations.

10. Explain in your own words why excessive precision and accuracy is highly undesirable, especially in manufacturing situations.

11. From a lab activity: $l = 24.3$cm and $F = 87.0$N. Using $\tau = Fl$, calculate τ, rounding appropriately.

12. From a lab activity: $m = 1.238$kg and $g = 9.81 \frac{m}{s^2}$. Using $w = mg$, calculate w, rounding appropriately.

13. From a lab activity: $\rho = 62.4 \frac{lb}{ft^3}$ and $h = 22$ ft. Using $P = \rho h$, calculate P, rounding appropriately.

14. From a lab activity: $R = 1027\Omega$ and $I = 5.23$mA. Using $V = IR$, calculate V, rounding appropriately.

15. From a lab activity: $F = 52.0 lb$ and $A = 2.00 in^2$. Using $P = \frac{F}{A}$, calculate P, rounding appropriately.

16. From a lab activity: $p = 0.0684 \frac{lb}{ft^2}$ and $\dot{V} = 0.050 \frac{ft^3}{s}$ Using $P = p\dot{V}$, calculate P, rounding appropriately.

17. From a lab activity: calculate and round appropriately 13.0mV – 1.0mV =

18. From a lab activity: calculate and round appropriately 17.1 cm - 9.51cm =

19. Calculate the percent error of activity that resulted in 9.52 when 9.81 was expected.

20. What is the percent error when 8.00 V was expected and the lab results were 8.88V?

Answers to R4 Student Exercises - Measurement

1. Accuracy: four significant digits, Precision: thousandths
2. Accuracy: three significant digits, Precision: ten-thousandths
3.
   a. A car with 8 cylinders goes 75 mph   8 is exact, 75mph is approximate
   b. Half of 7200 seconds. "Half" is exact, 7200 sec is approximate
   c. A calculator has 34 keys, and its battery lasted for 34 hours. 34 keys is exact, 34 hours is approximate
   d. Forty-five students drove 150 miles. Forty-five students is exact, 150 miles is approximate
4.
   a. 30.1 or 30.0   same accuracy, both have three significant digits
   b. 0.042 or 3450   3450 is more accurate
   c. 7000 or 0.000023   0.000023 is more accurate
   d. 128.0 or 128   128.0 is more accurate
5.
   a. 16.02 or 0.03   same precision, both precise to hundredths
   b. 25 or 20   25 is more precise
   c. 25 or $2\overline{0}$ same precision, both precise to ones
   d. 0.020 or 0.02   0.020 is more precise
6.
   a. 150,340 $15\overline{0},000$
   b. 0.02356198   0.0236
   c. 0.0449   already three significant digits, 0.0449
   d. 1.00000000   1.00
   e. 0.6999   0.700
7.
   a. 51.2 (tenths)
   b. 12.20 (hundredths)
   c. 62.1 (three significant digits)
   d. 0.270 (three significant digits)
   e. 2000 (only one significant digit in the given 2, ambiguous)
8. 46 (two significant digits in the current measurement)
9. In order for the customer specifications to be met, measurements must be made and recorded that maintain the required precision and accuracy, otherwise parts won't satisfy the requirements. Otherwise, customer specifications are lost.
10. Excessively accurate or precise measurements that exceed the customer requirements waste time and money, making the part more expensive to make than it needs to be.
11. 2110 N-cm
12. 12.1 N
13. 1400 $\frac{ft-lb}{s}$
14. 5370 mV
15. 26.0 $\frac{lb}{in^2}$
16. 0.0034$\frac{ft-lb}{s}$
17. 12.0mV
18. 7.6cm
19. 3.0%
20. 11.0%

Student Name(s) _____

_____

_____

_____

LAB OBJECTIVE

In this exercise students will identify an unknown metal by determining its density and comparing this value to a table of known densities.

Students will take measurements with a ruler and triple-beam balance, recording these measurements to their appropriate precision and accuracy. Students will then perform density calculations with these approximate numbers, rounding the result to the appropriate precision and accuracy.

Students will then compare their result to a table of known densities, calculating their percent error to the appropriate precision and accuracy.

Finally, students will answer questions regarding precision and accuracy as related to the measurements and calculations in this exercise.

FORMULAE USED HERE

Volume of a circular cylinder: $V = \pi r^2 h$

Volume of a block: $V = LWD$

Density: $\rho = \dfrac{m}{V}$

Table of Known Densities

Gold	19.3	gm/cm^3
Lead	11.3	gm/cm^3
Silver	10.5	gm/cm^3
Copper	8.9	gm/cm^3
Brass	8.6	gm/cm^3
Steel	7.8	gm/cm^3
Aluminum	2.7	gm/cm^3

APPARATUS

Triple-beam balance or digital scale

English ruler

Metric ruler or meter stick

Vernier caliper (optional)

Metal sample of known density

PROCEDURE

1. Measure the appropriate dimensions of the provided object in centimeters with the metric ruler, **estimating to the nearest half of the smallest division on the rule.** Record the measurements here in an organized fashion.

2. Calculate the volume of the object using an appropriate formula. Write the formula, numerical substitutions, the result and its units here. Round appropriately.

3. Use the triple-beam balance to measure the mass of the object in grams. Record the measurement here with appropriate precision and accuracy.

4. Calculate the density of the object using the density formula. Write the formula, numerical substitutions, the result and its units here. Round appropriately.

5. Identify the metal by comparing their measure/calculated value to the table of known densities provided above.

6. Calculate the percent error.
$$\% \text{ error} = \frac{|\text{actual} - \text{experimental}|}{\text{actual}} =$$

QUESTIONS

1. What is the smallest metric place value that can be read with a meter stick (or standard metric rule)?

2. The instructor discussed how an extra digit is often "squeezed out" of a tool by estimating distance between lines. Explain how this is done.

3. What is meant by **precision**?

4. What is meant by **accuracy**?

5. What is meant by a **direct measurement**? An **indirect measurement**?

6. The final result of the density indirect measurement has precision and accuracy. What is the precision and accuracy of your resulting density value?

7. Why is this final indirect measurement limited to this particular precision and accuracy?

8. A student makes two measurements, the current I = 4.31 A and E = 16.23 V. The student then uses the formula E = IR to calculate R. Manipulating the formula and making the numerical substitutions, the student gets…

    $$R = \frac{E}{I} = \frac{16.23V}{4.31A} = 3.7656612\Omega$$

    Explain to the student why this answer is wrong. What's the correct value?

A great many kinds of applied problems are solved using triangles. To name only a few, determining distances, navigation, structural design, astronomy, acoustics, force analysis, electric circuits, and velocities can be done by solving triangles. Allyn J. Washington, author of the widely used textbook, *Basic Technical Mathematics with Calculus* writes, "Because trigonometry has a great many applications in many areas of study, it is considered one of the most practical branches of mathematics." Thus we come to the study of trigonometry, the literal meaning of which is "triangle measurement".

Angle

An **angle** has three parts: the initial side, the terminal side, and the vertex. The vertex is the side where the two sides meet and also the point about which the two sides rotate. The size of the angle depends on the amount of rotation between the initial and terminal sides. When the rotation of the terminal side from the initial side is counterclockwise, the angle is positive. If the rotation is clockwise, the angle is negative. Also, it is possible for the terminal side to go completely around more than once.

A "degree" is the unit of measurement used to define the size of an angle. There are 360 degrees in one complete rotation of the terminal side, a complete circle. Units of degrees are often symbolized with a small circle in the superscript position of the number. For example, 90 degrees = 90°.

The initial and terminal sides form an angle at the vertex. This angle is often called the "reference" angle because, as we shall see later, its location establishes other names, ratios, and functions.

Right Triangle

Up until this point we've only dealt with two sides forming one angle, an initial side, a rotated terminal side, and the reference angle these two lines form at their connecting point, the vertex. But triangles of course have three sides and three angles, hence the "tri" in triangle. This third line connects the initial and terminal sides opposite the vertex. A box sometimes indicates the "right" or ninety degree angle of the right triangle.

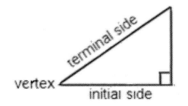

Naming the Sides of a Right Triangle

We usually start a trig problem by first placing the triangle in standard position. This means we place the triangle on a two dimensional graph with the vertex at the origin and the initial side on the x-axis. Any point along the terminal side can be represented by a coordinate (*x,y*). This creates a right triangle. A right triangle is one that has a "right", 90 degree angle. One side stands upright, perpendicular to the other. This angle is often shown with a small box in its corner.

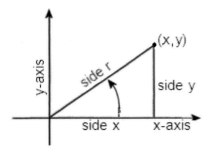

The first step in understanding trigonometry is to learn to name the three sides and three angles. One way is to name the initial side "*x*" because it lies on the x-axis. The vertical side would be called "*y*" because it lies on the y-axis. The terminal side is then either called "*r*" for "resultant". (More about resultants later.)

Sometimes, instead of naming the sides *x, y,* and *r*, we name them *a, b,* and c, with c always corresponding to the longest side, the terminal side *r*.

There is one more way of naming the sides, a way which may be best overall. In this case the initial side is called the "adjacent" side because it is adjacent the reference angle (one of the sides forming the reference angle). The vertical side is called "opposite" because it is opposite the reference angle. Finally, the terminal side is called the "hypotenuse". The hypotenuse is always the longest side of the three sides of a triangle. An important concept will be utilized and emphasized later: although the hypotenuse is always the longest side, which side we name "opposite" and "adjacent" *depends on the location of the reference angle.* That is, the opposite side is always opposite the reference angle and the adjacent side is always adjacent to the reference angle.

The sides that form the right angle are commonly called the "legs" of the triangle. Depending on how we might choose to name the triangle sides, the legs would then be *x* and *y, a* and *b,* or the opposite and adjacent sides.

Naming the Angles in a Triangle

When the sides of a triangle are labeled with lower case letters *a*, *b*, and *c*, the angles are labeled with capital letters *A, B*, and *C*.

Oftentimes, the angles are labeled with Greek letters. The Greek letters most often used for triangles are lower case θ (theta), α (alpha), and β (beta).

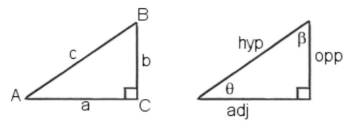

The Sum of the Angles in a Triangle

No matter what shape the triangle is and no matter what lengths of its sides, there will always be one very useful fact: *the sum of the three angles within the triangle will always be 180 degrees.* This is true for any triangle, not just right triangles. We use this to find the value of an angle when the other two are known.

When the angles are labeled *A, B*, and *C*, we have $A + B + C = 180$.

Note that since the right angle is always ninety degrees, the remaining two angles must add up to ninety degrees. When using the above naming conventions we have:

> **The Sum of the Angles**
>
> $A + B + C = 180$
>
> $A + B = 90°$ and $\theta + \beta = 90°$

It is important to understand that the angles may be given many different names, but the sum of the angles in any triangle, regardless of the angle nomenclature, always add up to 180°, and in a right triangle, the less-than-ninety angles always add up to 90°.

Example R5-1

Using the angle labels shown, find α when θ equals 20°.

Solution: $\alpha + \theta = 90°$

$\alpha = 90 - \theta$

$\alpha = 90 - 20$

$\alpha = 70$

Example R5-2:

From the diagram shown, find the missing angle. The triangle is not a right triangle.

Solution: $30 + 80 + \beta = 180$

$\beta = 180 - 30 - 80$

$\beta = 70°$

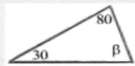

The Pythagorean Theorem

The sum-of-the-angles relationship is useful when working with only the angles, but it doesn't help us when solving for side lengths. We need another "tool" to work with sides. When two sides are known, we can use the Pythagorean Theorem (Pythagoras, 500 BCE) to calculate the remaining third side. This relationship is true for right triangles only.

The Pythagorean Theorem states that the sum of the squares of the "legs" of a right triangle equals the square of the remaining hypotenuse. Given any two sides of the right triangle, the remaining side can be found using this formula. For each of our three naming conventions:

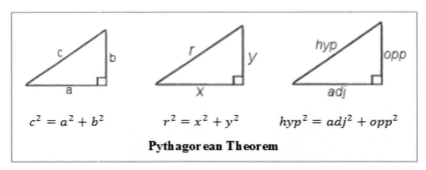

$$c^2 = a^2 + b^2 \qquad r^2 = x^2 + y^2 \qquad hyp^2 = adj^2 + opp^2$$

Pythagorean Theorem

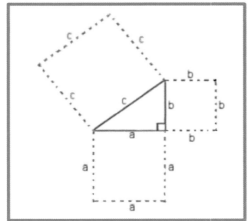

Example R5-3

Find c when $a = 3$ and $b = 4$.

Solution: $c^2 = a^2 + b^2$

$c^2 = 3^2 + 4^2$

$c^2 = 9 + 16$

$c^2 = 25$

$c = \sqrt{25}$

$c = 5$

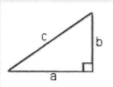

Example R5-4

Find the y when $x = 12.0$ and $r = 14.4$.

Solution: $r^2 = x^2 + y^2$

$$y^2 = r^2 - x^2$$
$$y^2 = 14.4^2 - 12.0^2$$
$$y^2 = 63.4$$
$$y = \sqrt{63.4}$$
$$y = 7.96$$

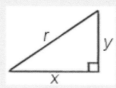

Example R5-5

Find the adjacent side when the opposite side equals 63.2 and the hypotenuse equals 66.8.

Solution: $hyp^2 = adj^2 + opp^2$

$$adj^2 = hyp^2 - opp^2$$
$$adj^2 = 66.8^2 - 63.2^2$$
$$adj^2 = 468$$
$$adj = \sqrt{468}$$
$$adj = 21.6$$

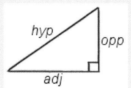

Pythagoras (~570BC – 495BC)

Pythagoras made influential contributions to philosophy and religion in the late 6th century BC. He is often revered as a great mathematician, mystic, and scientist and is best known for the Pythagorean theorem which bears his name.

It was said that he was the first man to call himself a philosopher, or lover of wisdom, and Pythagorean ideas exercised a marked influence on Plato, and through him, all of Western philosophy.

Tangent of a Right Triangle

What is a trig function? A trig function is a comparison of two sides, a ratio. We divide the value of one side by another. The result of this division, this comparison, is a trig function of the reference angle. The particular trig function depends on which sides are divided.

The value of the angles depends on the lengths of the sides, that is, the angles are a function of the lengths of the sides.

Since there are three sides, there are six ways of dividing one side by another, but three are reciprocals of the other three. As a result, only three trig functions are of particular use when solving right triangles. Each one of these trig ratios, or functions, has a name. The first one we will study is the tangent (tan) of an angle.

If (x,y) is any point along the terminal side of an angle, then $\frac{y}{x}$ is the tangent (tan) of that angle. "Tangent" is the name given to this particular ratio of sides. It is the ratio of the "legs". When the sides are labeled opposite, adjacent, and hypotenuse, the ratio is the same but written $\frac{opp}{adj}$.

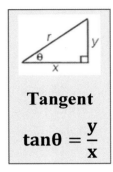

Tangent

$$\tan\theta = \frac{y}{x}$$

Note how the opposite side is opposite the reference angle and the adjacent side is adjacent to it. We name the sides opposite or adjacent depending on the location of the reference angle. This is why we write the function $\tan\theta$, with the angle designation included. This tangent ratio is the function of a particular angle in the triangle.

A simple yet extremely important concept to grasp here is that the value of the angle depends on the lengths of the sides (the angles are a function of the lengths of the sides) and that any particular tangent ratio has a particular corresponding angle.

Let us first calculate tangent ratios.

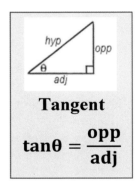

Tangent

$$\tan\theta = \frac{opp}{adj}$$

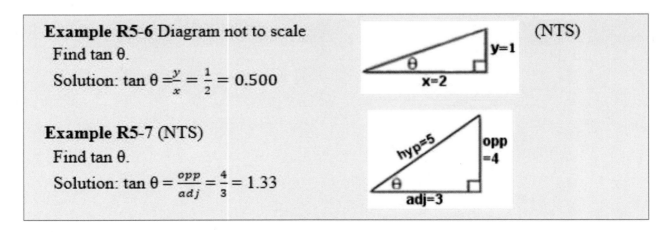

Example R5-6 Diagram not to scale (NTS)

Find tan θ.

Solution: $\tan\theta = \frac{y}{x} = \frac{1}{2} = 0.500$

Example R5-7 (NTS)

Find tan θ.

Solution: $\tan\theta = \frac{opp}{adj} = \frac{4}{3} = 1.33$

48

Example R5-8 (NTS)

Find tan θ.

Solution: Here we define which sides are opposite and adjacent by the location of the reference angle θ. This reference angle establishes the names. The 6.2 is adjacent to θ while the 8.8 is opposite θ. So *adj* = 6.2 and *opp* = 8.8. Of course the 10.8 is the hypotenuse, the longest side, but this side is not part of the tangent ratio.

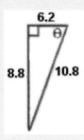

$$\tan \theta = \frac{opp}{adj} = \frac{8.8}{6.2} = 1.42$$

Example R5-9 (NTS)

Find tan θ.

Solution: Here the opposite and adjacent sides are equal.

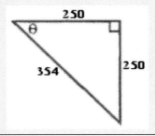

$$\tan \theta = \frac{250}{250} = 1.00$$

(By inspection, θ = 45°)

Now that we have found the tangent of a reference angle by calculating the ratio of the opposite and adjacent sides, we can now find the value of the angle. Long ago, long lists of trig functions were tabulated so that someone could calculate the ratio of sides then *look up the value of the angle in the table.* Today we tend to use our calculators, but let us first begin by using the Table of Trigonometric Functions on the following page.

In Example R5-6, tanθ = 0.500. In other words the opposite side is half that of the adjacent side. This defines the shape of the triangle, including the value of the angles. If this triangle were to be then drawn to scale, the angle can then be measured directly with a protractor and listed in a table next to 0.500 under the heading "tangent". In this manner, the trig ratios and corresponding angles of many different triangles can then be listed and used to solve triangles.

The trig table in the appendix is not an especially precise one. Note how the value of the angle is precise only to the nearest degree. Either we *interpolate* between values or use much longer tables with more entries, but this one will serve our purpose. Given tan θ = 0.500, let us look up the corresponding angle in the table. Under the Tangent heading find the number closest to 0.5000. This would be 0.5095. Now look to the left under the Angle heading and read the angle. 27° lines up with the horizontal row. So, when the tangent ratio is 0.5000, the reference angle is approximately 27°.

It is important to learn to write these functions properly, depending on whether we're solving for the angle or the ratio. If tan θ = 0.500 then θ = arctan 0.500 = 27°

Although the *ratio* of the sides is called *the function,* the *angles* are found using an *inverse function,* written as *arctangent.*

But since the advent of electronic calculators, $\tan^{-1}$ has been often used instead of arctangent simply because the six letter arctan won't fit on the calculator key. It's OK to write it this way, but the technically correct way to right it is arctan, finding the arc, or angle. It should always be spoken "arctan", along with "arcsine" and "arccosine".

If $\tan \theta = 0.500$ then $\theta = \arctan 0.500 = 27°$ or $\theta = \tan^{-1} 0.500 = 27°$.

Example R5-10

Use both the trig table and your scientific calculator to find θ when $\tan \theta = 0.8333$.

Solution: $\theta = \arctan 0.833 = 39.8°$

Example R5-11

Use both the trig table and your scientific calculator to find θ when $\tan \theta = 2.769$.

Solution: $\theta = \arctan 2.769 = 70.1°$

Example R5-12

Use both the trig table and your scientific calculator to find θ when $\tan \theta = 1.000$.

Solution: $\theta = \arctan 1.000 = 45.0°$

Example R5-13

Use both the trig table and your scientific calculator to find β when $\tan \beta = 0.3333$.

Solution: $\theta = \arctan 0.3333 = 18.4°$

(Note that when tangent θ equals one, the legs are equal, so the angle is 45°. When tangent θ is less than one the angle is less than 45°. When tangent θ is greater than one the angle is greater than 45°.)

Sine and Cosine of a Right Triangle

The two remaining trig functions (ratios) we will study are sine (sin) and cosine (cos). They work the same way as the tangent function except that different sides are used to make the ratios. Unlike the tangent function, these ratios use the *hypotenuse* side of the triangle.

Remember that the hypotenuse might go by other names; the "r" or "terminal" side.

Written in terms of *opp, adj,* and *hyp*:

Sine	Cosine
$\sin\theta = \dfrac{\text{opp}}{\text{hyp}}$	$\cos\theta = \dfrac{\text{adj}}{\text{hyp}}$

Written in terms of x, y, and r:

Sine	Cosine
$\sin\theta = \dfrac{y}{r}$	$\cos\theta = \dfrac{x}{r}$

Also, remember that while the *hypotenuse* is always the longest side, we name the two remaining sides by their position with respect to the reference angle. Note that in the two triangles below, the *opposite* side is opposite the reference and the *adjacent* side is adjacent to the reference angle. In other words, we name the side "opposite" or "adjacent" depending on the location of the reference angle. This is the best way to name sides because triangles can be located and oriented in various ways in any given application, but the *opposite* side is always opposite the reference angle, hence the name "*opposite*".

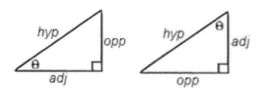

Every triangle has six parts, three sides and three angles. If any three parts are known (but including at least one side to establish scale) the triangle can be completely solved. In a right triangle, one of the angles is known, ninety degrees, so if two additional parts are known (again including at least one side) the rest of the triangle can be solved. In the following examples see how we can now know any two parts about a right triangle (including at least one side) and solve for the rest.

Example R5-14

Solve for y in the right triangle. (NTS)

Solution: We choose the sine function because it is the ratio that contains the known side and the side to be found.

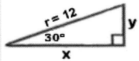

$$sin\theta = \frac{y}{r}$$

$$sin30° = \frac{y}{12} \text{ (From the trig table } sin30° = 0.500.)$$

$$y = 12sin30° = 12(0.500)$$

$$y = 6.00$$

Example R5-15

Calculate the adjacent side. (NTS)

Solution: We choose cosine because it is the ratio that contains the known side, hypotenuse, and the side we which to find, the adjacent.

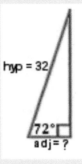

$$cos\theta = \frac{adj}{hyp}$$

$$cos72° = \frac{adj}{32}$$

$$adj = 32cos72° = 32(0.309)$$

$$adj = 9.90$$

Example R5-16

Calculate the hypotenuse, side "r".

Solution: The given side (8.00) is adjacent to the given angle.

$$adj = 8.00$$

We choose the cosine function because it contains the known side and the side to be found.

$$cos\theta = \frac{adj}{hyp}$$

$$cos46° = \frac{8}{r}$$

$$r = \frac{8}{cos46°} \text{ (Careful of the formula manipulation!)}$$

$$r = \frac{8}{0.695}$$

$$r = 11.5$$

General Approach to Solving Right Triangles

When solving a right triangle we are given two parts of the triangle, either an angle and a side or two sides. At least one side must be known or we cannot determine the overall scale of the triangle. We choose the appropriate trig function by looking at which parts are given together with which parts are to be found. In simple terms, each function contains three parts, so choose the function that has the two parts you know. Then use this function to solve for the remaining unknown. Any given triangle could be solved in different ways, so there is no universal step-by-step procedure we should always follow. We have options.

As described earlier, we have three "weapons" at our disposal:

- Three trig functions (Right triangles only.)

$$sin\theta = \frac{opp}{hyp}$$

$$cos\theta = \frac{adj}{hyp}$$

$$tan\theta = \frac{opp}{adj}$$

- Pythagorean Theorem (Right triangles only.)

$$c^2 = a^2 + b^2$$

- Sum-of-the-angles = 180° (Any shaped triangle.)

$$\Sigma A = 180° \quad \text{(Capital sigma } \Sigma \text{ means "sum of".)}$$

The standard triangle with the given names (shown at right) is used in the following examples. (NTS)

Example R5-17

Find x, y, and β when r = 120ft and θ = 16°.

Solution: There are various options.

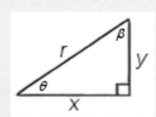

Finding y first using the sine function…

$$sin\theta = \frac{y}{r}$$

$$y = rsin\theta = 120sin16° = 120(0.2756)$$

$$y = 33.1$$

Finding x using the cosine function…

$$cos\theta = \frac{x}{r}$$

$$x = rcos\theta = 120cos16° = 120(0.9613)$$

$$x = 115$$

Using Pythagoras to check the sides…

$$r = \sqrt{115^2 + 33.1^2} = 120 \checkmark$$

Finding angle β…

$$\beta = 90 - 16 = 74°$$

Example R5-18

Find θ, β, and y when $r = 25.0$ and $x = 8.00$.

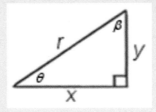

Solution: There are options.

Finding y using Pythagoras…
$$y = \sqrt{r^2 - x^2} = \sqrt{25^2 - 8^2}$$
$$y = 23.7$$

Finding θ using cosine…
$$\cos\theta = \frac{x}{r}$$
$$\theta = \arccos\frac{8}{25} = arccos\ 0.320 \quad (\text{or } cos^{-1}0.320)$$
$$\theta = 71.3°$$

Finding β…
$$\beta = 90 - 71.3 = 18.7°$$

Checking with tangent…
$$\theta = tan^{-1}\frac{y}{x} = tan^{-1}\frac{23.7}{8.00} = 71.3°\sqrt{}$$

Example R5-19

Find r, θ, and β when $x = 35$ and $y = 35$.

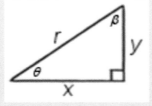

Solution: As always, there are options.

By inspection $\theta = 45°$, but proving using tangent…
$$\tan\theta = \frac{y}{x} = \frac{35}{35} = 1$$
$$\theta = arctan 1 = 45°\sqrt{}$$

By inspection $\beta = 45°$, but proving using ΣA with one $90°$ angle…
$$\beta = 90 - 45 = 45°$$

Finding r using Pythagoras…
$$r = \sqrt{x^2 + y^2} = \sqrt{35^2 + 35^2}$$
$$r = 49.5$$

Checking r using sine…
$$\sin\theta = \frac{y}{r}$$
$$r = \frac{y}{\sin\theta} = \frac{35}{\sin 45°} = \frac{35}{0.7071}$$
$$r = 49.5\sqrt{}$$

Example R5-20

Find θ, r, and x when $\beta = 68°$ and $y = 21.5$.

Solution: Options.

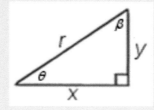

Using θ to keep y opposite and x adjacent...

$$\theta = 90 - 68 = 22°$$

Using tangent to find x...

$$tan\theta = \frac{y}{x}$$

$$x = \frac{y}{tan\theta} = \frac{21.5}{tan22°} = \frac{21.5}{0.404}$$

$$x = 53.2$$

Finding r using Pythagoras...

$$r = \sqrt{x^2 + y^2} = \sqrt{53.2^2 + 21.5^2}$$

$$r = 57.4$$

Checking with sine...

$$sin\theta = \frac{y}{r}$$

$$sin22° = \frac{21.5}{57.4} = 0.375 \checkmark$$

55

Use the labeling system shown in the diagram to solve for the remaining unknowns in each problem. Write a solution. Round to three significant digits.

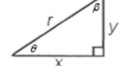

1. Given $y = 4.00$, $x = 7.00$, and $r = 8.06$
   find $\sin \theta$, $\cos \theta$, and $\tan \theta$.

2. Given $\theta = 50.0°$ and $x = 6.70$
   find y, r, and β.

3. Given $x = 56.8$ and $r = 79.5$
   find y, θ, and β.

4. Given $\theta = 32.0°$ and $r = 56.8$
   find β, x, and y.

5. Given $\theta = 70.0°$ and $y = 140$
   find β, x, and r.

6. Given β = 12.0° and r = 18.0
find θ, x, and y.

7. Given x = 86.7 and r = 167
find θ, β, and y.

8. Given y =150 and r = 345
find β, θ, and x.

9. Given θ = 77.8° and y = 6700
find β, x, and r.

10. Given x = 30.0 and r = 50.0
find θ, β, and y.

Answers to Solving Right Triangles Exercises
1. sinθ = 0.496, cosθ = 0.868, tanθ = 0.571 **2.** y = 7.99, r = 10.4, β = 40° **3.** θ = 44.4°, β = 45.6°, y = 55.6 **4.** y = 30.1, β = 58.0°, x = 48.2 **5.** r = 148, β = 20.0°, x = 51.0 **6.** θ = 78.0°, x = 3.70, y = 17.6 **7.** θ = 58.7°, β = 31.3°, y = 143 **8.** θ = 25.8°, β = 64.2°, x = 311 **9.** β = 12.2°, x = 1450, r = 6850 **10.** y = 40.0, θ = 53.1°, β = 36.9°

R5 Student Exercises - Generic Applications of Right Triangles

1. How tall is the utility pole? (Ans: 21.8 feet)

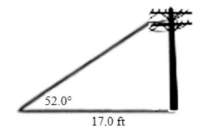

52.0°

17.0 ft

2. How far away (horizontal distance) is the flagpole? (Ans: 22.5 feet)

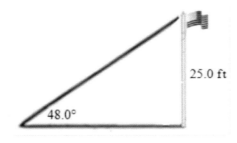

25.0 ft

48.0°

3. What is the altitude (height) of the missile? (Ans: 89,200 meters)

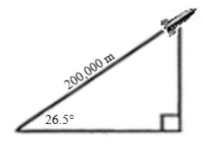

200,000 m

26.5°

4. How tall is the antenna? (12.3 feet)

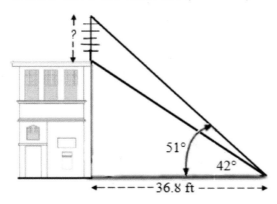

5. How long is the shoreline on the lakefront property? (Ans: 306 feet)

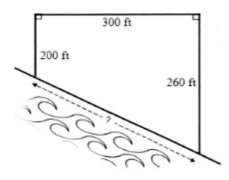

6. How tall is the pyramid? (Ans: 234 feet)

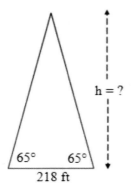

Chapter 1   DISPLACEMENT

Each energy system has its own version of a displacement quantity. In general, the **displacement quantity** describes the amount of motion or change that occurs in the particular energy system.

In mechanical systems the displacement is a "how far" type of quantity. **Translational mechanical systems** are those involving the motion of objects such as cars, pool balls, and bullets, objects that move more or less in a straight line. **Rotational mechanical systems** involve spinning objects. In translational mechanical systems the displacement is the **distance**, or how far something moves, while in rotational mechanical systems the displacement is an **angle**, or how far an object has turned.

In the remaining three energy systems however, the displacement is "how much". In fluid systems the displacement is the **volume** of fluid moved, in electrical systems the displacement is the how much **charge** moved, and in thermal systems the displacement is the amount of **heat** moved. Displacement is the term describing all of these quantities in general.

Each of these five displacement quantities has its own terminology, literal and systems of measurement units. The purpose of this chapter is to study these conventions.

Unifying Principle of Displacement:

Each energy system has a quantity that describes the amount of motion or change that occurs in that system. This quantity is called the displacement quantity.

Displacement

Distance (d) *in Translational Mechanical Systems*
Angle (Ɵ) *in Rotational Mechanical Systems*
Volume (V) *in Fluid Systems*
Charge (q) *in Electrical Systems*
Heat (Q) *in Thermal Systems*

Most people are familiar with distance. Distance is how far we might commute to school every day, how far an object may have fallen, the height of a mountain, or the length or width of a room in your home. Note that there are many different types of distance that go by different names depending on the application ("h" for height, "L" for length, "W" for width, "ℓ" for lever arm), but each is in general called a form of **displacement**. Usually, lower case letter "**d**" represents distance in formulae, unless otherwise noted.

Although there are many different units of measurement for various quantities that might be used depending on the particular applications, each quantity has what is called a fundamental unit, a unit that is most commonly used in formulae and numerical calculations. In the English system of measurement, a variety of old and interesting units have been handed down to us over the ages. Units of distance such as "rods" are not commonly used these days. Certainly the reader is familiar with inches, feet, yards, and miles. But the fundamental unit of measurement for distance is **feet**.

In metric SI, the fundamental unit of measurement for distance is the **meter**, which is a little more than a yard. Powers of ten are then used to create small or large units based on the meter, with a centimeter (cm) being about as wide as the tip of your little finger and one kilometer (km) roughly two-thirds of a mile.

Distance Units

1 meter = 1000mm = 100 cm = 0.001 km = 39.37 inch = 3.281 ft
1 foot = 0.3048 meter = 12 in
1 mile = 5280 ft = 1.609km = 1609m

Distance verses Translational Displacement

Translational mechanical systems include quantities where objects such as cars and bullets are moving about. How far they move is called their translational displacement, or their translational distance. Here we now examine the distinction between translational displacement and translational distance.

Distance is a scalar quantity having only magnitude whereas translational displacement is a vector quantity having both magnitude and direction.

Say you leave your home to take a walk, going to the park and back. The distance you walked might be measured by the total number of steps you took, all the way over and all the way back, regardless of the twists and turns you made. But translational displacement includes all those twists and turns, so your translational displacement is zero. You're back home. You're ultimately right back where you started, no displacement.

Trigonometry is used to calculate translational displacement. We will limit discussion here to right triangles. For example, say a jogger runs three miles toward the east, then turns north and runs an additional four miles. Calculate the joggers distance and also her translational displacement.

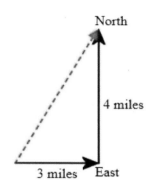

These types of problems are solved with vectors and vector addition. Earlier we learned that vectors have two parts, magnitude and direction. **Vectors** are represented graphically as arrows.

The arrow's length represents the **magnitude** of the vector. Unless making rough sketches, the length of the arrow is drawn to scale, say each centimeter representing a particular distance in miles, just like on a scaled map. (Centimeter measure scale is preferred over inches because centimeters are decimal.)

The **direction** of the vector is its angle. Here we set the system up in standard form, where east is equivalent to zero degrees (to the right) and north is equivalent to ninety degrees (upward). The vectors to be added are drawn **tip-to-tail**. This creates a triangle which can then be solved. The red, dotted vector is drawn from where the jogger started to where she stopped, a straight line from start to stop "as the crow flies". This is her displacement, how far she ended up from where she started. This displacement is the hypotenuse of our right triangle.

Calculating her distance is a simple direct sum; she ran 3 mi + 4 mi = 7 miles. Her distance is 7 miles, d = 7 mi.

But her translational displacement is measured from where she started to directly where she ended up, a straight line "as the crow flies". Using the triangle we have created, this displacement can now be solved in either or both of two ways. The first is the **graphical solution**. If a scale diagram is drawn well enough, the displacement magnitude and direction can be measured directly from the triangle with a rule and protractor. The second method is the **mathematical method**. With the use of a roughly drawn triangle, the displacement magnitude and direction can be calculated mathematically using trigonometry. Typically the mathematical method is preferred because scale diagrams are tedious and time consuming to draw. Also, graphical solutions are always very approximate, limited to the precision of the drawing and skill of the person making the sketch. This is why vector additions are typically solved by drawing only rough sketches, not scale diagrams, then applying the math. But scale diagrams do indeed provide a useful check to the mathematical solution.

A mathematical solution for our jogger's displacement magnitude and direction is shown. Since this is a right triangle, we can use the Pythagorean Theorem to calculate the translational displacement (the hypotenuse). The arctangent function is used to calculate the vector angle. All vector solutions must have two parts, the magnitude and the direction.

Example 1-1

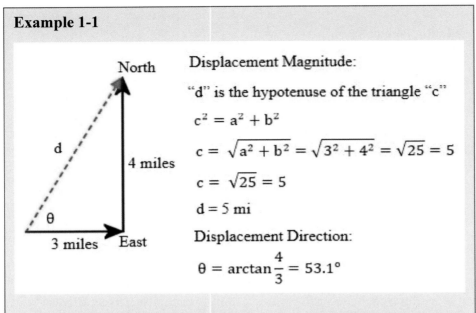

North

Displacement Magnitude:

"d" is the hypotenuse of the triangle "c"

$$c^2 = a^2 + b^2$$

$$c = \sqrt{a^2 + b^2} = \sqrt{3^2 + 4^2} = \sqrt{25} = 5$$

$$c = \sqrt{25} = 5$$

$$d = 5 \text{ mi}$$

Displacement Direction:

$$\theta = \arctan\frac{4}{3} = 53.1°$$

d

4 miles

θ

3 miles   East

The jogger's displacement is a vector quantity having both magnitude and direction.

The jogger's translational displacement is d = 5 miles at 53.1° (NE).

Note that both translational distance and translational displacement are represented by the same literal, d. This is something that is not usually troublesome, but we must still be cautious understanding which is which and what is being solved for in applied problems.

There are situation when it doesn't matter, where the translational distance and displacement are equal anyway. For example, if the jogger in the above example never took any turns and just ran in a straight line, then her distance equals here displacement. Another example might be a top fuel dragster at a race track. The race track is a straight quarter-mile length. Here it can be said the distance and displacement of a race car is equal.

Other times it is impossible to make the distinction between the two so general displacement must be assumed. For example, someone in a vehicle taking a cross-country trip from Maine to California would take roads that curve and turn repeatedly, but while the car's odometer could provide an accurate distance after the trip is completed, a map can only provide a rough displacement in advance of the trip.

In physics triangles are usually drawn in "standard form" or "standard orientation". Further explained on page 68, the x axis is drawn horizontally while the y axis is drawn vertically, perpendicular to the x axis. Zero degrees is positioned to the right, with angles increasing in the counter-clockwise direction, 90° at the top, 180° at the left, and 270° at the bottom, or downward. Unlike orienteering conventions, for compass directions we will position east at 0°, north at 90°, west at 180°, and south at 270°.

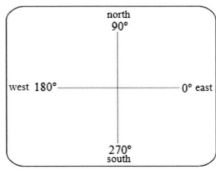

1. The distance between Presque Isle and Portland is roughly 290 miles. What is this distance in kilometers?

2. Convert 1.5 meters to millimeters.

3. Houlton, Smyrna Mills, and Bridgewater form an approximate right triangle on the map. Bridgewater is about 20 miles north of Houlton and Smyrna Mills is about 15 miles west of Houlton. Calculate the distance between Smyrna Mills and Bridgewater.

4. A fighter jet leaves base and flies 145 kilometers south, then turns rather sharply to the east and continues for another 65 kilometers. The pilot sees she has only enough fuel for another 150 kilometers. **a.** What heading (direction) should the pilot fly if she were to fly directly to base? **b.** Does she have enough fuel to return to base?

5. When squaring foundations, sometimes contractors calculate the diagonal distance; the two diagonals will match when the structure is square. What should be the diagonal distance of a foundation 68 feet long and 48 feet wide?

6. A group of soldiers left camp and walked 6 km to the west, turned south, then walked another 8 km. Calculate **a.** their distance and **b.** their displacement from camp.

Answers: **1.** 467 km **2.** 1500 mm **3.** 25 mi **4. a.** 114° (or 65.9° NW) **b.** No. 159 km away
5. Approximately 83' $2\frac{1}{2}$" **6. a.** 14km **b.** 10km at 233.1° (or 53.1° SW)

1.2 Displacement in Rotational Mechanical Systems   ANGLE (θ)

In rotational mechanical systems objects turn. Gears, crankshafts, flywheels, fans, motors, bolts and screws all turn or rotate.

In rotational systems the displacement is how far an object turns, an angle. The angle is generally represented as angle theta (θ) in formulae. There are a number of different units of measurement that might be used to measure angles. All of these units of measurement are neither English nor SI and are used in both systems. These include degrees (°), revolutions (rev), and **radian measure** (rad). Our focus here in this section is to understand radian measure. There are formulae in which the angle *must* be in radian measure.

Angles

What makes radian measure so useful is that it is a dimensionless ratio. A ratio is a type of fraction that is a comparison of two values that have the same dimensions, or units. For example, the rise/run comparison of a set of stairs is a ratio because the rise in inches is compared to the run in inches. When two values having the same units are divided, the units cancel, producing a number that is "dimensionless". A dimensionless value can be inserted into calculations without affecting any of the other units.

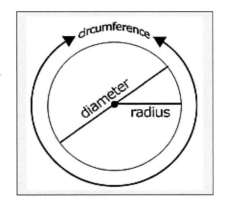

To begin to understand what radian measure is, let's review the various parts of a circle, their names, and their relationships.

The **circumference** (c) of a circle is the total length around, a special word for the perimeter. The **diameter** (d) of the circle is the circle's width at its widest, and the **radius** (r) is half the diameter. It should be noted that "d" is used for distance but also diameter. Care should be taken as to which quantity is involved.

We will learn that it's not a coincidence that "radius" and "radian" are nearly the same word.

Circles are divided into 360 parts called degrees, the degree symbol being "°". So the circumference encloses an angle of 360°.

Angles are typically drawn in what is called "standard position". A horizontal line is drawn through the center of the circle. This is commonly called the "x axis". The zero degree point is the right-most end of this horizontal line. Then a line perpendicular to the x-axis is drawn vertically through the center of the circle. This is commonly called the "y-axis".

A line from the center of the circle to the right-most edge of the circle is of radius length. This line is called the **initial side** of an angle. Angles are typically measured starting at this line, the initial point. Then imagine rotating this line, keeping its left-hand end at the center, or **vertex**, to create an angle. This new line is called the **terminal side**. The angle created is the rotational displacement between the initial and terminal sides.

When the initial side rotates in a counter-clockwise direction, the angle is given a positive sign, say +60°.

When the initial side rotates in a clockwise direction, the angle is given a negative sign, say -30°. Negative thirty degrees equals positive 330 degrees.

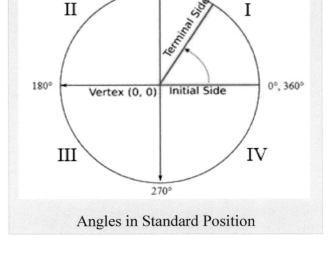

Angles in Standard Position

Angles can exceed 360°, making many revolutions.

The complete circle is typically divided into four sections called quadrants. These are numbered using Roman numerals. A **quadrant** is a circular sector of 90°. Quadrant I is between 0° and 90°, II between 90° and 180°, III between 180° and 270°, and IV between 270° and 360°. Sometimes large angles are expressed as **acute** angles (angles less than 90°) positioned on the circle by quadrant number and measured from horizontal. For example, 30° in quadrant III would be equal to an angle of 210° measured from 0°.

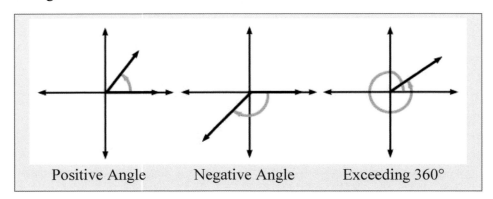

Positive Angle   Negative Angle   Exceeding 360°

A well-known relationship regarding circles is π, the ratio of the circumference to the diameter, a number approximated to 3.14. In other words, in all circles the circumference is always 3.14 times its diameter. Since the diameter is twice the radius, d = 2r, then π can also be defined as the ratio of the circumference to twice the radius. *So the circumference is 2π times the radius.*

$$\pi = \frac{c}{d} \qquad \pi = \frac{c}{2r}$$

$$c = 2\pi r$$

Radian Measure

An angle of one radian is created when the arc length equals the radius. When the terminal side rotates such that the length of the curved outer part of the circle, the "crust" of our piece of pie, equals the radius, the angle is one radian. Radian measure is then a dimensionless ratio, the ratio of the length of the arc to the length of the radius with units cancelling.

Since c = 2πr, that is the circumference is 2π times more than the radius, it takes 2π, or 6.28, radii lengths to go around once, one revolution, 360°. "Rad" appears after the number of the angle, but it is not a unit of measurement per se, but the name of the dimensionless ratio. It is used only to indicate how the angle is measured, and does not affect any dimensional analysis.

Solving physics problems often involves converting units. In rotational systems there are some formulae that will only work if the angle is in radian measure so, if say revolutions is given, the angle must be converted to radians.

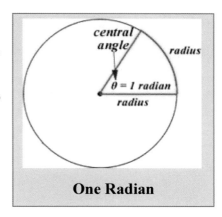

One Radian

Angle Units

1 rev = 360° = 2π rad

1 rad = 57.3°

Example 1-2
Convert 1000 revolutions to radians.

Solution:

$$1000\cancel{\text{rev}}\left(\frac{2\pi\text{rad}}{1\cancel{\text{rev}}}\right) = 6280 \text{ rad}$$

Example 1-3
Convert 150⁰ to radians

Solution:

$$150^o\left(\frac{1\text{ rad}}{57.3^0}\right) = 2.62 \text{ rad}$$

Example 1-4
A wheel with a diameter of one meter roles three revolutions. What distance did it roll?

Solution:
The wheel covers three times its circumference.
$$c = 2\pi r \quad d = 3(2\pi r) = 3(6.28)(0.5\text{m}) = 9.42 \text{ m}$$

Radian measure is based on the ratio of the arc length to the radius. This is represented by the formula

$$\theta = \frac{s}{r}$$

where θ represents the angle in radians, s the arc length, and r the radius. To say it another way, radian measure is a comparison of the arc length to the radius. The units of s and r must be the same and must cancel to leave a dimensionless ratio we call radians.

1. In which quadrant is 220°?

2. In which quadrant is 2.18 radians?

3. How is the radian ratio defined?

4. Convert 1200 revolutions to radian measure.

5. Convert 72° to radian measure.

6. Calculate the circumference of a circle that has an eighteen inch radius.

7. A wheel has a radius of fifteen inches. What distance does it travel if it rolls twenty revolutions?

8. Calculate the angle of a circular sector having an arc length of one foot and a radius of three inches.

9. Calculate the radius of a circular sector having an angle of 42.0° and an arc length of sixty feet.

10. A roundabout with a radius of ninety feet was constructed at a busy street intersection. What is the distance of one lap around?

Answers: 1. III **2.** II **3.** Radian measure is the ratio of the arc length to the radius.
4. 7540 rad, **5.** 1.26 rad **6.** 113 in **7.** 157 ft **8.** 4 rad **9.** 81.9 ft **10.** 565 ft

1.3 Displacement in Fluid Systems   VOLUME (V)

Fluid systems move fluids. Water pumps provide us water to drink and bathe. Oil pumps drive pistons to lift heavy weights. Fluids carry away heat in cooling systems.

A **fluid** can be either a **liquid** or a **gas**. Liquids and gasses can be treated quite nearly the same way except for one major distinction: liquids are virtually incompressible, meaning it takes incredible pressure to change its volume. Gasses on the other hand are easily compressed and change volume with little pressure. A **hydraulic system** is a fluid system that uses a liquid, like water or oil. For example, hydraulic cylinders on a tractor implement are operated by a hydraulic pump that pushes hydraulic oil in and out of pistons and other actuators. A **pneumatic system** is a fluid system that uses a gas, such as air, as the operating fluid. Many useful pneumatic tools such as finish nailers and air wrenches are driven by compressed air. Whether hydraulic or pneumatic, the quantity displaced in these systems is a volume of liquid or a volume of gas. The displacement quantity in fluid systems is the **volume** of fluid.

Volume

All matter takes up space. The three-dimensional space a substance occupies is its **volume**.

The displacement quantity in fluid systems is the volume, or the amount of fluid. There are numerous units of measurement for fluid volume, including fluid measure in gallons and liters and also units of length cubed, such as cubic feet or cubic centimeters.

There are times when we will need to calculate the volume, given the geometric shape of the vessel containing the fluid. We will study three of these; the cube, a rectangular solid (the shape of a brick), and the right circular cylinder (like in pistons).

Volume of Three Geometric Shapes

A **cube** is a three-dimensional solid object that has six facets, or sides. The volume of the cube is found by multiplying the three sides. Since each side is the same, this value has an exponent of three. (Any power of three is often said to be "cubed" even when the calculation has nothing to do with the volume of a cube.)

As we cube the numbers, we also cube the units. For example, a cube 2 inches x 2 inches x 2 inches has volume of 8 cubic inches, usually written 8 in³. So when calculating the volume of these shapes, the units of measurement for, say, length, width, and height, must all be the same.

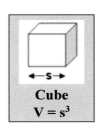

Cube
$V = s^3$

A **rectangular solid** is similar to a cube except that all three sides aren't the same. The volume is calculated in a similar way though, multiplying each of its three sides together. Here the sides are labeled length (L), width (W), and height (h). Given this labeling scheme the volume formula is V = LWh

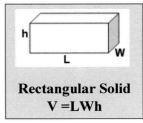

Rectangular Solid
V =LWh

A **right circular cylinder** is shaped like a pipe. It is called a "right" circular cylinder because it is usually drawn standing upright. It has a round end. The volume of a piston is a right circular cylinder. The volume is found by calculating the circular area of the end using A =πr² then multiplying this value to the height. So the volume of the right circular cylinder is V =πr²h.

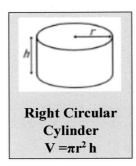

Right Circular Cylinder
V =πr² h

Cubic units of any length are legitimate units of volume for all substances, including fluids.

Fluid Measure

In the English system the fundamental units of volume is the **gallon** (gal). It is surprising how many gallons will fit into one cubic foot (1 ft³), 7.48 gallons.

The English system has always been a cumbersome hodgepodge of units. There are pins, quarts, teaspoons and tablespoons, drams, ponies, jiggers, and jacks. These have a long and interesting history of development, but have been troublesome, especially in medical applications. The metric system solved all these problems by establishing the liter. It's *so* simple it's confusing!

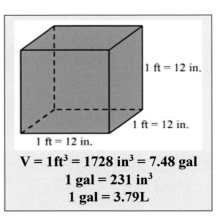

1 ft = 12 in.
1 ft = 12 in.
1 ft = 12 in.

V = 1ft³ = 1728 in³ = 7.48 gal
1 gal = 231 in³
1 gal = 3.79L

A **liter** (L) is the space of a cube 10 centimeters on a side. The volume of this cube is V = (10cm)(10cm)(10cm) = 1000cm³. Typically cubic centimeters is abbreviated cm³ when solids are involved, but cc for liquids, but these measures are the same, cubic centimeters, and can be used interchangeably.

Since milli means thousandths, then there are also 1000 milliliters, or 1000 mL, in one liter.

There are 3.79 liters in one gallon.

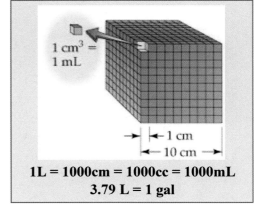

1 cm³ = 1 mL

1 cm
10 cm

1L = 1000cm = 1000cc = 1000mL
3.79 L = 1 gal

Mass and Weight as a Measure of How Much

Although this section of our displacement chapter focuses on volume as the displacement quantity in fluid systems, there are other quantities that are a measure of an amount of fluid, or how much of any type of substance. Here it is necessary to introduce two quantities different from volume but useful as a measure of how much: mass and weight.

Mass (m) is the total amount of matter, or material, of a substance. Be careful to understand that the volume, or size, of the substance is irrelevant. For example, the amount of mass in a room of air can be equal to the mass of a small chunk of metal. The chunk of metal is much smaller than the room of air but the amount of mass is equal. As far as mass is concerned, their size doesn't matter. There is also an important distinction between mass and weight. Mass and weight are different things. **Weight** (w) is the force created when gravity acts on the mass, pulling it down. So while a particular object would weigh less on the Moon than on Earth, the mass is the same. An object could even be weightless, but it still has mass.

In the English system of units, a long history of different standards of measurement combined to give us what we have today. It is interesting that the concept of mass, a quantity independent of gravity and more fundamental than weight, was supposedly never grasped until Isaac Newton in the 1600's. Even British money came to be in pounds because in those days people bartered and traded by the weight of things, pounds. So the old English system never had a unit of mass until relatively recently. But an English unit of mass was finally adopted. In the English system of units mass is measured in **slugs** (no abbreviation). Other than in isolated applications like water treatment facilities, the slug is seldom used. One slug is a fairly large unit, equal to a weight of 32.2 pounds in Earth's gravity, but more on that later.

> Ever wonder about the strange abbreviation for pounds, lb? Its roots go way back to the goddess Libra, who purportedly established a system of weights and measures for people in ancient times.

In the SI metric system of units, mass is measured in grams (g). The metric system makes things easy: the size of one **gram** was established by calling the mass of one cubic centimeter of water one gram. As a result, for water, $1 \text{ cm}^3 = 1cc = 1mL = 1g$. So we have mass measured in grams, but what about metric weight units? The SI unit of weight is the **Newton** (N), named after Sir Isaac Newton.

Later we will study the concept of density, where different materials will have different amount of mass packed into a particular volume, like the air and chuck of metal mentioned above. But remember, when it comes to just the quantities of mass and weight, size doesn't matter.

> **Metric System**
> mass (m) in grams (g)
> weight (w) in Newtons (N)
>
> **English System**
> mass (m) in slugs (slugs)
> weight (w) in pounds (lb)

The Gallon

Most people are familiar with the gallon, like a gallon of milk. But there are three different types of gallon measures we must aware of: the US fluid gallon, the US dry gallon, and the UK gallon. Each of these is a different size.

It might be best to compare the three different gallon measures in cubic inches.

> **1 US fluid gallon = 231 in^3**
>
> **1 US dry gallon = 269 in^3**
>
> **1 UK gallon = 278 in^3**

The fluid gallon is of course associated with fluids. Remember fluids include both liquids and gasses. For example, for proper ventilation rates, the volume of air in a room would be measured in fluid gallons. Another example is the "displacement" of an engine, which is the total volume of the cylinders of the engine, and since a fuel vapor is involved, engine displacement is measured in fluid volume. (A popular and successful Buick engine is the 3.8 L, sometimes called the "3800". Its total cylinder volume is 231 in^3, or one fluid gallon.)

The dry gallon is used when non-fluid materials are measured, such as flour, sand, and aggregate.

Perform the following calculations, rounding to three significant digits.

1. Calculate the volume of a cube that is 2.20 inches on a side.

2. Calculate the volume of cylinder that is 135 mm long and has a diameter of 24 mm.

3. Convert 12.0 dry gallons to cubic inches.

4. Convert 12.0 fluid gallons to cubic inches.

5. How many cubic centimeters in two liters?

6. How many milliliters in two liters?

7. A high efficiency urinal saves 87% of water usage at 0.125 GPF (gallons per flush). Calculate the Lpf (liters per flush) of the system.

8. Explain how mass and weight are not the same things.

9. How many grams of water are there in two liters?

10. What is the mass of 1210 cm^3 of water?

11. For health and safety reasons, the volume of air in a room must be completely ventilated within a limited amount of time. Calculate the volume of air of a room having dimensions: W = 24ft, L = 56ft, H = 10ft.

12. The six cylinder car engine is a GM 231 in^3. What is the displacement (volume) of each cylinder in fluid gallons?

Answers: 1. 10.6 in^3 **2.** 61,$\overline{0}$00 mm^3 (or 61.0 cm^3) **3.** 3230 in^3 **4.** 2770 in^3 **5.** 20$\overline{0}$0 cm^3 **6.** 20$\overline{0}$0 mL **7.** 0.474 Lpf **8.** Mass is the amount of material contained in an object while weight is the force created when gravity acts on the mass **9.** 20$\overline{0}$0 g **10.** 1210 g **11.** 13,400 ft^3 **12.** Since one gallon equals 231 in^3, one of the six cylinders is one sixth of a gallon, or 0.167 gal

1.4 Displacement in Electrical Systems   **CHARGE (q)**

Electrical systems generate and use electricity. Users, or **electric loads**, such as light bulbs, televisions, ovens, elevators, computers, water pumps, phones, radio stations, automobiles, and countless machines are everywhere in our homes, schools, businesses, and shops and factories. Needless to say, electrical systems play an ever more important part of every aspect of our lives.

Electrons and the Atomic Theory

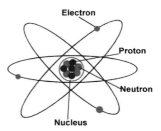

Atomic theory is the idea that matter is made up of smaller parts called **atoms**. The atoms are then made of even smaller parts. **Protons** and **neutrons** cling together in the center, or **nucleus**, while a swarm of much smaller particles, **electrons**, spin around the nucleus in a great cloud.

Protons are defined as having positive charge while the electrons are defined as being negatively charged. An atom is balanced when it has an equal number of protons and electrons, equal and balanced charge, but if there is an unbalance in their number, the atom seeks to either get rid of the extra electrons or to replace a missing ones. This state of imbalance is called **charge** (q). The number of missing or extra electrons is the measure of the amount of charge.

Materials that have few electrons (and/or also electrons that are positioned closely to the nucleus) do not easily allow electrons to flow, so are good electrical **insulators**. On the other hand, materials that have many electrons are good **conductors** of electrons. Indeed, some materials like copper have many **free electrons** that loosely and randomly float about from atom to atom. It is these free electrons in conductors that are easy to get moving. The amount of electrons that are moved is the displacement in electrical systems, the charge.

Electrons are very small, so many, many billions of electrons can be displaced. A convenient unit of measurement for charge was established. The measurement unit of charge is the **coulomb** (c). One coulomb is 6.25×10^{18} electrons!

The coulomb can best be understood by comparing electrical systems to fluid systems. When we displace water, we don't count the individual molecules of water that were moved, we measure this in gallons or liters. In much the same way, in electrical systems the displacement isn't the number of individual electrons, but a larger measure of many electrons, the coulomb.

> Why this particular amount of electrons?
>
> $1c = 6.25 \times 10^{18} \, e^-$
>
> This question will be examined in our subsequent chapters on work, energy, and power. We will soon see that establishing this particular number of electrons for one unit of charge makes for useful and interchangeable energy and power calculations.

Charles-Augustin de Coulomb (1736 –1806)

Coulomb was a French physicist and military Captain. He was best known for developing Coulomb's law, the definition of the electrostatic force of attraction and repulsion, but also did important work on friction, torsion, and other phenomena. The SI unit of electric charge, the coulomb, was named after him.

Charge verses Current

It is important to understand that charge, measured in coulombs, is a completely different quantity than current, measured in amperes. Think of charge as stationary electrons, say in a battery that is not connected to anything, while current is the *flow rate* at which the electrons move when the battery discharges to run a device. If we compare these electrical units to fluid systems, the charge is analogous to stationary fluid measured in gallons while the current is the flow rate of the fluid measured in gallons per minute. One ampere equals one coulomb per second.

The Limitations of the Electron Theory

The atomic electron theory is an old, popular, and useful explanation for electricity, but it is very simplistic and limited. The electron theory only works for particular metals, and they must be solid metals (not in liquid state). The model does not work for when water conducts electricity. For example, no electrons flow through the human body when shocked. It also fails to explain most other electrical phenomena including plasmas (like your television screen), lightning, aurora, and fluorescence. But this does not mean the electron theory hasn't any use or value; the electron theory works very well when applied to electric currents, voltage rises and

drops, and resistances in electronic circuits where solid conductors at limited temperatures are used.

1. Describe the three basic components of an atom.

2. Explain what is meant by one coulomb.

3. How many electrons are there in 120 coulombs?

4. Using electron theory, explain why certain materials conduct electricity and others don't.

5. What is the displacement quantity in electrical systems?

Answers: 1. The atom is made up of proton and neutrons at the center, or nucleus, while electrons circle outside of the nucleus. **2.** One coulomb is a specific number of electrons, $1c = 6.25 \times 10^{18} \, e^-$ **3.** $7.50 \times 10^{20} \, e^-$ **4.** Conductors have many free electrons that can easily be displaced while insulators have few electrons and even fewer free electrons **5.** The displacement quantity in electrical systems is the charge measured in coulombs.

1.5 Displacement in Thermal Systems   HEAT (Q)

Heat (Q)

Every substance has a special relationship between the temperature change it experiences and the heat that causes the temperature change. Temperature difference is a quantity we will investigate further later, but for now it is only important to understand that when heat is added to or taken away, substances undergo temperature changes. *Be advised: temperature and heat are different things and the words/quantities cannot be used interchangeably.*

Temperature change is represented by the letters ΔT. These two letters represent one quantity. The delta (Δ) means "change in" or "difference". Any change or difference is calculated by subtracting. Here T_f represents the "final temperature" and T_i the "initial temperature". Their difference is the temperature change the substance experiences. For

$$\Delta T = T_f - T_i$$

example, if some water is heated from 70°F to 100°F, the temperature change is 30F°. Note that the units of measurement are °F for temperature but F° for temperature difference.

If you touch a hot stove, the energy enters your hand because the stove is warmer than your hand. There is a temperature difference. When you touch a piece of ice, however, energy transfers from your hand into the colder ice. Once again, a temperature difference. The direction of energy transfer is always from the warmer matter to the neighboring cooler matter. The energy transferred from one object to another because of a temperature difference between them is called **heat**. Heat is a form of energy found in the vibrating motion of atoms and molecules.

Although temperature is a measure of the amount of heat of a substance, it is important to understand that heat and temperature are not the same thing. One way to think of this is to think of temperature as the way a particular substance responds or reacts to heat. We will soon see how different materials have different relationships between heat and temperature.

Specific Heat (c)

Ever notice that while you have to use an oven mitt to remove a hot roasting pan from a hot oven, you can touch the aluminum foil in the roasting pan without burning yourself? Both the metal of the roasting pan and the aluminum are the same temperature, but each requires a different amount of heat to be at that temperature. As a result there is less of a ΔT between your fingers and the foil than between your fingers and the metal roasting pan. This relationship between the heat "contained" in the substance and the temperature of the substance is called its **specific heat** (c), or specific heat storing capacity.

Different substances have different capacities for storing internal heat energy. The heat storing capacities of a substances are compared to water.

> The **specific heat** is defined as the amount of heat required to produce a particular change in temperature in a particular amount of a particular substance.

British Thermal Unit

In the English system, the unit of measurement for heat is the **British Thermal Unit** or **Btu**. The Btu was established by using water as the standard, the relationship between heat and temperature of water. One Btu is defined as the amount of heat required to cause one pound of water to change temperature 1F°. The specific heat of water is 1 Btu/lb·F°.

> One Btu is the amount of heat required to change the temperature of one pound of water 1 F°.
> Specific heat of water: c = 1 Btu/lb·F°
> Specific heat of ice: c = 0.49 Btu/lb·F°

Calorie

In the metric system, the unit of measurement for heat is the **calorie** (cal). Once again water is used as the standard. One calorie is defined as the amount of heat required to cause one gram of water to change temperature by 1C°.

> One calorie is the amount of heat required to change the temperature of one gram of water by 1C°.
>
> Specific heat of water: c = 1 cal/g·C°
> Specific heat of ice: c = 0. 49 cal/g·C°

Changes of State

Substances can exist in four different forms, called the **states of matter**: solid, liquid, gas, and plasma. We will study three; solid, liquid, and gas. These states of the material are thermal processes, changing when the substance reaches certain temperatures. Water freezes/thaws at 32°F or 0°C. This temperature is called the **melting point** of a substance. Water begins boiling at 212°F or 100°C under normal pressures. This of course is called the **boiling point** of a substance. Other substances have different melting and boiling point temperatures.

> **ICE/WATER/WATER VAPOR**
> **Freezes/thaws: T = 32°F = 0°C**
> **Boils/condenses: T = 212°F = 100°C**

Calculating Sensible Heat (Q) and Latent Heat (Q_f or Q_v)

Sensible heat is the amount of heat that causes a substance to undergo temperature changes *between phase changes* while **latent heat** is the heat needed *during phase changes*.

The reason phase change heat is called "latent", or hidden, is that during phase changes there is no change in temperature.

Think of what happens when you put a pot of water on the stove. Sensible heat gets the water up to boiling temperature. But once it begins to boil the water remains at 212°F (or 100°C). It remains at this temperature until all the liquid is converted to a gas.

Three factors affect sensible heat; the amount of substance (m), the specific heat of the substance (c), and the temperature change (ΔT). The formula for sensible heat is $Q = mc\Delta T$.

> **Sensible Heat**
> $$Q = mc\Delta T$$

Example 1-5
How much heat is needed to bring 500 grams of water from an initial temperature of 20°C to boiling temperature?
Solution:

$\Delta T = 100 - 20 = 80C°$

Using the sensible heat formula we have

$$Q = mc\Delta T = 500g \left(1 \ \frac{cal}{g \cdot C°}\right)(80C°) = 40,000cal \ or \ 40kcal$$

Note how the units cancel to result in calories.

The English system of units must be treated a little differently. In most situations, units of mass must be substituted in for literal "m" in any formula, grams in the metric system and slugs in the English system. But here we have an exception. Remember that one Btu was defined by establishing one *pound* of water changing one degree. So for thermal formulae, pounds must be substituted for m in the heat formulae. Another issue is that water is typically given in gallons, not pounds, so units of gallons must be converted to pounds.

> **1 gallon = 8.34 lbs**

Example 1-6
How much heat is needed to bring 2 gallons of water from 70°F to boiling temperature?
Solution:

$\Delta T = 212 - 70 = 142F°$

"m" $= 2gal \left(\frac{8.34 \ lb}{1gal}\right) = 16.7 \ lb$

Using the sensible heat formula we have

$$Q = mc\Delta T = 16.7lb \left(1\frac{Btu}{lb \cdot F°}\right)(142F°) = 2370 \ Btu$$

Pounds cancel, the English standard for the Btu, with the Btu remaining in the answer.

During phase changes when a substance changes between the states of solid, liquid, and gas, there is not temperature change. Heat must be continually added or removed in order for a complete phase change, but the temperature remains the same. Phase changes occur at different temperatures for different substances, but we will limit discussion here to water.

Only two factors affect latent heat, the amount of the substance (m) and a heat rate. **Heat of fusion** (H_f) the amount of heat a substance needs per gram (or per pound) to freeze or melt and **heat of vaporization** (H_v) is the amount of heat need per gram (or per pound) for the substance to boil away.

The formulae for latent heat is simply the product of the mass and the heat rate constant.

Water	**Water**
Latent Heat of Fusion $Q = mH_f$ **Latent Heat of Vaporization $Q = mH_v$**	$H_f = 144$ **Btu/lb = 79.8 cal/g** $H_v = 970$ **Btu/lb = 540 cal/g**

Example 1-7
How much heat is needed to completely boil away one liter of water already at 100°C?
Solution:

   1L water = 1000 cc = 1000g
   The water is already at the boiling point temperature.

$$Q = mH_v = 1000g \left(540 \frac{cal}{g}\right) = 540,000 \text{ cal or } 540 \text{ kcal}$$

Example 1-8
How much heat is needed to completely boil away 1.5 gallons of water that is initially at 70°F?
Solution:

   The total heat added is the sensible heat bringing the water from 70° to 212° plus the
   latent heat required to vaporize it once it reaches the boiling point.

$$\text{"m"} = 1.5 \text{ gal}\left(\frac{8.34 \text{ lb}}{1 \text{ gal}}\right) = 12.5 \text{ lb}$$

$$\Delta T = 212 - 70 = 142 \text{ F}°$$

$$Q_T = mc\Delta T + mH_v = 12.5 lb\left(1 \frac{Btu}{lb \cdot F°}\right)(142 F°) + 12.5 lb\left(970 \frac{Btu}{lb}\right)$$

$$Q_T = 1780 + 12,130 = 13910 \text{ Btu}$$

The total amount of heat added to or removed from a substance can also be the sum of the sensible and latent heats.

Please round the numerical answers to three significant digits.

1. One "spring" day in Aroostook County, Maine, the daytime high was 52°F but the nighttime low was −11°F. Calculate the temperature change.

2. Explain what is meant by the specific heat of a substance.

3. An electric water heater has a 38 gallon tank. How much heat is needed to raise the temperature of the water in the tank from 72°F to 165°F?

4. How much heat is needed to completely boil away 10 gallons of water already at the boiling point?

5. How much heat is needed to completely boil away 10 gallons of water that is initially at 70°F?

6. How much heat is needed to raise the temperature of 20 liters of water from 26°C to 86°C?

7. How much heat is needed to completely boil away 10 liters of water that is already at the boiling point?

8. How much heat is needed to completely boil away 10 liters of water initially at 28°C?

9. How much heat is needed to melt 1kg of ice initially at −15°C?

Answers: 1. 63.0F° **2.** The specific heat of a substance is the heat storing capacity of the substance. Also, the relationship between the heat in the substance and the temperature of the substance. **3.** 29,500 Btu **4.** 45,$\overline{0}$00 Btu **5.** 199,000 Btu **6.** 12$\overline{0}$0 kcal **7.** 54$\overline{0}$0 kcal **8.** 6120 kcal **9.** 87.2 kcal

Chapter 2 RATE

Technicians of all kinds measure and control a wide variety of rates in a variety of interacting energy systems so that the systems operate efficiently and safely.

In this chapter we will study the different types of rates technicians encounter. In general, **rate** is a quantity that describes how rapidly an occurrence takes place. Here we will apply a unifying principle for rate to each of our five energy systems. In every case, rate is a comparison of the displacement in the energy system to the elapsed time.

> ## Unifying Principle - Rate
>
> $$\text{RATE} = \frac{\text{DISPLACEMENT QUANTITY}}{\text{ELAPSED TIME}}$$

2.1 Rate in Translational Mechanical Systems

Speed and Velocity

The most fundamental rate in a mechanical system is speed. Following our unifying principle of displacement over time, speed is the distance divided by time. Speed is a scalar quantity, the rate at which an object moves without regard for direction. The displacement quantity is the distance the object moved. The elapsed time is the amount of time the object took to move this distance. Velocity, on the other hand, is a vector quantity where direction is considered.

$$\text{SPEED} = \frac{\text{DISTANCE}}{\text{ELAPSED TIME}}$$

$$v = \frac{d}{t}$$

Note that the two formulae for speed and velocity are the same, using lower case "v" for speed or velocity, lower case "d" for distance or translational displacement, and lower case "t" for time. (As always, make the distinction between upper and lower case letters.) As mentioned earlier during discussion on the distinction between distance and translational displacement, using "d" for both distance and displacement is just something we must be conscious of, understanding which we are dealing with depending on the application.

$$\textbf{VELOCITY} = \frac{\textbf{TRANSLATIONAL DISPLACEMENT}}{\textbf{ELAPSED TIME}}$$

$$\textbf{v} = \frac{d}{t}$$

In section 1.1, *Distance in Translational Mechanical Systems*, we studied an example involving a jogger who ran three miles east then turned north to run an additional four miles. We drew a vector diagram and calculated both her distance and her translational displacement. Let's expand on this example by introducing an amount of time into the scenario, and calculating the jogger's speed and velocity.

Let's say the jogger took 42 minutes to complete the run, a distance of seven miles but a displacement of five miles. Calculate the jogger's speed and velocity in miles per hour.

Example 2-1

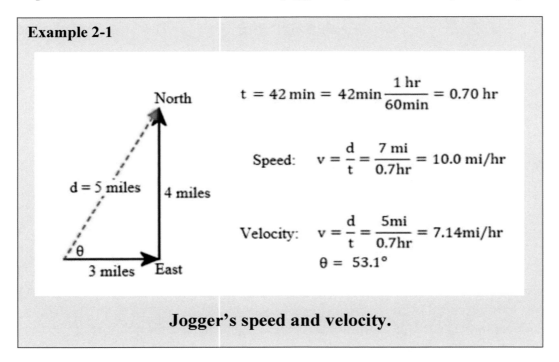

$$t = 42\,min = 42min\,\frac{1\,hr}{60min} = 0.70\,hr$$

$$\text{Speed:} \quad v = \frac{d}{t} = \frac{7\,mi}{0.7hr} = 10.0\,mi/hr$$

$$\text{Velocity:} \quad v = \frac{d}{t} = \frac{5mi}{0.7hr} = 7.14mi/hr$$

$$\theta = 53.1°$$

Jogger's speed and velocity.

The formulae for rate has three quantities: the displacement, the time, and the rate. This formula can be used to calculate any one of these quantities given any two of the three variables. When the rate is given, here speed or velocity, we must be prepared to manipulate the formula for the unknown displacement or time, and substitute the given values to calculate the unknown value. We must also be careful that the units are consistent. Here are a few illustrative examples.

Example 2-2

How long will it take to drive a distance of 560 miles at an average speed of 52mph?

d = 560 mi, v = 52 mph

$$v = \frac{d}{t}$$

$$t = \frac{d}{v} = \frac{560mi}{52mph} = 10.8 \text{ hrs}$$

Note how the units cancel, inverting mi/hr and multiplying to cancel miles. Also, a good rule-of-thumb is that the displacement is always divided by the rate to calculate time.

Acceleration

The above translational examples involve the average speed or velocity of an object. The speed is constant throughout the elapsed time. But the speed or velocity can change, too. Objects can speed up or slow down. The rate of change of the speed or velocity is called translational acceleration.

Acceleration (a) is the rate of change of the velocity where "a" is the acceleration, "v_f" is the final velocity, and "v_i" is the initial velocity. The change in velocity is the difference between the final and initial velocities. This is commonly written as Δv, pronounced "delta v", meaning "change in velocity.

$$\text{ACCELERATION} = \frac{\text{CHANGE IN VELOCITY}}{\text{ELAPSED TIME}}$$

If we divide speed by time, the result is a rate of a rate. The units of acceleration are units of velocity divided by time. Dividing the standard units of velocity in the English system, ft/sec, by time we get

$$\frac{ft/sec}{sec}.$$

Sometimes the expression "feet per second per second" is used. If you think about it, it really means what it says: the velocity in feet per second is changing per second.

Acceleration

$$a = \frac{v_f - v_i}{t}$$

$$\Delta v = v_f - v_i$$

$$a = \frac{\Delta v}{t}$$

Remember the rules for dividing fractions? When we divide compound fractions, we invert the denominator and multiply. Doing this we get

$$\frac{ft}{sec} \cdot \frac{1}{sec} = \frac{ft}{sec^2}.$$

All this works the same way in SI where we have $\frac{m/sec}{sec} = \frac{m}{sec^2}$.

Various combinations of units can be used for acceleration, but the most common standards are ft/s^2 in the English system and m/s^2 in SI.

The expression "deceleration" is commonly used. But in the sciences this expression is seldom used. "Acceleration" is the term used whether an object is slowing down or speeding up. The difference is mathematical; deceleration is negative acceleration. So don't be confused if you are asked to calculate the "acceleration" when clearly the object is slowing.

Example 2-3

A driver increases the speed of a car uniformly from 30 mph to 50 mph in 10 seconds.
Calculate the acceleration of the car.

Solution A: $\quad a = \frac{\Delta v}{t} = \frac{50mph - 30mph}{10sec} = \frac{20mph}{10sec} = 2.0$ mph/sec

The car's velocity increased by 2.0 mph every second.
Although technically correct, these are not standard units of acceleration. Conversions need to be made to get ft/sec².

Solution B:

$v_f = 50$mph $= 73.3$ ft/sec

$v_i = 30$mph $= 44.0$ ft/sec

$\Delta v = 73.3$ ft/sec $- 44.0$ft/sec $= 29.3$ ft/sec

$a = \frac{\Delta v}{t} = \frac{29.3 \text{ ft/sec}}{10sec} = 2.93 \text{ft/sec}^2$

There are useful formulae that can be used in conjunction with $a = \frac{\Delta v}{t}$ to solve acceleration problems. The velocity $v = \frac{d}{t}$ can be used in acceleration problems, but we must remember it is the *average* velocity, total distance over total time.

Given an initial and final velocity in a problem involving uniform acceleration, the average velocity is just that, the average, and can be calculated like any average. Just add them up and divide by two.

Be careful to distinguish between the three different types of velocity in an acceleration problem, the initial velocity (v_i), the final velocity (v_f), and the average velocity (v_{avg}).

> **Average Velocity During Uniform Acceleration**
>
> $v_{avg} = \frac{v_f + v_i}{2}$

Example 2-4

An object accelerates from 10 ft/sec to 30 ft/sec in 5 seconds. Find
  a. the acceleration.
  b. the distance covered during the 5 second time interval.

Solution:

$$a = \frac{\Delta v}{t} = \frac{20 \text{ ft/sec}}{5 \text{ sec}} = 4 \text{ ft/s}^2$$

$$v_{avg} = \frac{v_f + v_i}{2} = \frac{30 \text{ft/s} + 10 \text{ft/s}}{2} = \frac{40 ft/s}{2} = 20 \text{ft/s}$$
(v_{avg} could have been done by inspection)

$v = d/t$
$d = vt = 20\text{ft/s}(5\text{s}) = 100\text{ft}$

Acceleration Due to Gravity

All the above formulae and strategies discussed above can be applied to falling objects. Treat these types of problems as an acceleration problem, one in which the acceleration is known.

Objects speed up when they fall. They accelerate. The Earth acts on the object pulling it down. This acceleration is called the **acceleration due to gravity**. Although this acceleration is constant, falling bodies will experience drag that slows them down a bit. Here we will neglect air drag.

A special literal is used to represent acceleration due to gravity, the letter "g".

> **Acceleration Due to Gravity**
>
> **g = 32.2 ft/sec² = 9.81 m/s²**

An interesting thing about falling objects is that the mass or weight does not affect the objects speed or acceleration; all objects regardless of their weight fall at the same rate, g. We will investigate the reason for this in an upcoming chapter.

Example 2-5

An object falls for 2.5 seconds. Find
   a. the velocity just before it hits ground.
   b. the distance it fell during this time interval.

Solution:

$a = g = 9.81 \text{m/s}^2$

$v_i = 0$

$a = \dfrac{v_f - v_i}{t} = \dfrac{v_f - 0}{t} = \dfrac{v_f}{t}$

$g = \dfrac{v_f}{t}$

$v_f = gt = 9.81 \text{m/s}^2 (2.5\text{s}) = 24.5 \text{ m/s}$

$V_{avg} = \dfrac{v_f + v_i}{2} = \dfrac{24.5 \text{m/s} + 0 \text{m/s}}{2} = \dfrac{24.5 m/s}{2} = 12.3 \text{m/s}$

$d = vt = 12.3 \text{m/s}(2.5\text{s}) = 30.8 \text{ meters}$

Velocity and Acceleration Formulae

$$v = \frac{d}{t} \qquad a = \frac{v_f - v_i}{t} \qquad V_{avg} = \frac{v_f + v_i}{2} \qquad a = \frac{2d}{t^2} \text{ when } v_i = 0$$

Example 2-6

An object is measured to take approximately 0.452 seconds to fall one meter. Calculate the object's acceleration.

Solution:

$v_i = 0$

$a = g = 9.81 \text{m/s}^2$

$a = \dfrac{2d}{t^2} = \dfrac{2(1\text{m})}{(0.452\text{s})^2} = 9.79 \text{ m/s}^2$

Of course the object falls at the acceleration due to gravity. More digits (accuracy) in the time measurement might have yielded an acceleration closer to the accepted g = 9.81 m/s².

ACCELERATION **L2**

Student Name(s) _____

_____

_____

Freefall: Acceleration Due to Gravity

Objectives

Upon finishing this lab, the student will be able to...

1. distinguish between average velocity, final velocity, and initial velocity.
2. define acceleration as it applies to freefalling bodies.
3. use velocity and acceleration formulae to calculate acceleration given the distance and time an object falls
4. experimentally determine acceleration due to gravity.
5. calculate percent experimental error.

Discussion

Part 1: The Basic Formulae

The most fundamental rate in mechanical systems is an object's *velocity*. Average velocity is calculated by dividing the total distance an object moves by the time during which the movement occurred. Although an object's velocity might change or vary, the object's *average velocity* is simply the total distance it moved divided by the time period:

$$v_{avg} = \frac{d}{t}$$

Where *v* is the average velocity, *d* is the distance covered by the object, and *t* is the time interval.

Acceleration is the rate at which the velocity changes (speeding up or slowing down), and is expressed by the formula

$$a = \frac{v_f - v_i}{t}$$

where *a* is the acceleration, v_f is the final velocity, v_i is the initial velocity, and *t* is the time.

When an object accelerates from rest $v_i = 0$ so the acceleration formula can be simplified to

$$a = \frac{v_f}{t} \quad \text{when } v_i = 0$$

Units of measurement for acceleration are usually ft/s^2 or m/s^2.

When an object accelerates it has an initial velocity and final velocity. Given these values, the average velocity can be calculated using a simple and useful average formula

$$v_{avg} = \frac{v_f + v_i}{2}.$$

When an object begins its movement from rest, the initial velocity is zero. In this case, the average velocity is simply half the final velocity.

$$v_{avg} = \frac{v_f}{2} \quad \text{and} \quad v_f = 2v_{avg} \quad \text{when } v_i = 0$$

Neglecting drag, all objects fall at the same acceleration due to gravity (g):

$$a = g = 9.81 \, m/_{s^2} = 32.2 \, ft/_{s^2}$$

Part 2: Building a Formula for *a* given *d* and t

In this activity, the student will experimentally determine the acceleration due to gravity on an object initially at rest, by measuring the time duration of the fall over a set distance. Here we will combine three formulae to derive one formula to solve for *a* given *d* and *t*.

$$v = d/t \; , \quad v_f = 2v_{avg}, \quad a = \frac{v_f}{t}$$

We begin by substituting $v_f = 2v_{avg}$ into $a = \frac{v_f}{t}$:

$$a = \frac{2v_{avg}}{t}$$

Since $v_{avg} = d/t$, we can substitute d/t into our acceleration formula:

$$a = \frac{2\left(d/t\right)}{t}$$

Simplified we have a formula by which we can calculate uniform acceleration given distance and time, but for $v_i = 0$ *only*:

$$a = \frac{2d}{t^2}$$

Equipment:

One photogate (timer stop)

One photogate timer

Glider/projectile release control

Glider/projectile electromagnet

Glider/projectile timer start cable

Meter stick

Metal Ball (ball bearing)

Your instructor may have other similar ball-drop apparatus that serves the same function

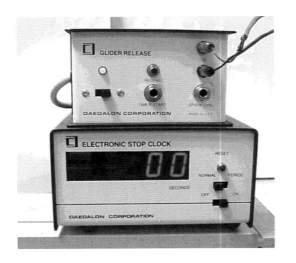 

Procedures:

1. Assemble the equipment as shown with the projectile release at the top and the stop photogate at the bottom. **Position the photogate at the convenient drop distance of one meter, d = 1m.**

2. Connect the leads from the top-mounted electromagnet to its release control.

3. Connect the photo gate to the timer. The electromagnet release control simultaneously releases the projectile and starts the clock. The projectile then stops the clock when it breaks the LED beam in the photogate.

4. Release the ball, timing the drop (or any convenient distance). Do at least three runs, repositioning the ball and readjusting the gate positions as needed. Record the run times in the table provided.

Data Table

Run	t (in seconds)	a (in meters/sec^2)	% error
1			
2			
3			
4			

Calculations

1. Using the formula derived above, $a = \dfrac{2d}{t^2}$, calculate the acceleration, given the drop distance and drop time. Record the values in the table.

2. Calculate percent error from expected given the known acceleration due to gravity:

$$\frac{|\text{expected} - \text{experimental}|}{\text{expected}} = \frac{|9.81 - a|}{9.81} \times 100$$

The absolute value bars in the numerator means to disregard ± sign.

Student Challenge

Given a = g and using $a = \dfrac{2d}{t^2}$, calculate the amount of time the object *should* have taken to fall one meter. (Hint: Solve for *t* in $a = \dfrac{2d}{t^2}$.)

2.1 STUDENT EXERCISES   TRANSLATIONAL RATE

1. What is the *unifying* equation for rate?

2. How does the unifying equation for rate apply to translational systems?

3. A conveyor belt moves an object twelve feet in three seconds. Calculate the translational velocity of the conveyor belt.

4. A commuter travels between Caribou and Presque Isle, a distance of about thirteen miles, in 15.5 minutes. Calculate the commuter's average velocity in mph.

5. An automobile accelerates from an initial velocity of 15 mph to 25 mph. Calculate the vehicle's average velocity.

6. Convert 55 mph to ft/s.

7. Convert 88 km/hr to m/s.

8. How long will it take for a trucker to travel from Maine to California, a distance of approximately 3500 miles, at an average velocity of 35 mph?

9. An object accelerates from 10 ft/sec to 25 ft/sec in 5 seconds. Calculate
   a. the acceleration.

   b. the average velocity.

   c. the distance the object covered during the acceleration.

10. An object falls for 3.5 seconds. Calculate the object's final velocity (just before it hits ground).

11. A stock car completes one lap of a circular track in 62.8 seconds. The track diameter is half a mile. Find the linear velocity of the stock car in mph.

12. A rock is dropped down into a well. Timing the fall, it takes 4.2 seconds from the time of its release to the time a watery splash is heard. Neglecting the time it takes for sound to travel up the well, calculate the depth of the well in meters.

13. A team of soldiers march five kilometers north, then turn east and march an additional 2.5 kilometers. Their march takes 5 hours. Calculate
   a. their speed.

   b. their velocity.

14. A "dragster" is a race car designed to compete on a quarter-mile track called a "drag strip". During a practice run, a dragster begins from a stop and accelerates to 180 mph. Calculate

   a. the dragster's average velocity in ft/s.

   b. the dragster's time on the quarter-mile track.

   c. The dragster's acceleration in ft/s^2.

Answers: 1. rate = displacement divided by time **2.** "distance" divided by time **3.** 4 ft/s
4. 50.3 mph **5.** 20.0 mph **6.** 80.7 ft/s **7.** 24.4 m/s **8.** 100 hrs or 4.17 days **9.a** 3 ft/s
9.b 17.5 ft/s **9.c** 87.5 ft **10.** 34.3 m/s or 113 ft/s **11.** 92.4 mph **12.** 86.5 m **13.a** 1.5 km/h
13.b 1.12 km/h **14.a** 132 ft/s **14.b** 10.0 sec **14.c** 26.4 ft/s^2

2.2   Rate in Rotational Mechanical Systems

Rate in rotational systems is very much like translational rate discussed above but for one major difference; instead of distance, the displacement is angular, **angular displacement**. The same terms are used except the word "angular" is placed in front to describe it as rotational.

Following our unifying principle of displacement over time, the **angular velocity** is the angular displacement divided by the time. The **angular acceleration** is the rate at which the angular velocity changes.

$$\text{ANGULAR VELOCITY} = \frac{\text{ANGULAR DISPLACEMENT}}{\text{ELAPSED TIME}}$$

$$\text{ANGULAR ACCELERATION} = \frac{\text{CHANGE IN ANGULAR VELOCITY}}{\text{ELAPSED TIME}}$$

Conceptually, the rotational formulae are the same as the translational formulae, except that the displacements quantity is angular and Greek letters are used. Note how the translational formulae are essentially the same as the rotational formulae.

While "d" is used to represent displacement (or distance) in translational systems, theta (θ) is used to represent angular displacement in rotational systems.

While "v" is used to represent velocity (or speed) in translational systems, lower case omega (ω) is used to represent angular velocity in rotational systems.

While "a" is used to represent acceleration in translational systems, lower case alpha (α) is used to represent angular acceleration in rotational systems.

Analogous Velocity and Acceleration Formulae

Translational	Rotational
$v = \dfrac{d}{t}$	$\omega = \dfrac{\theta}{t}$
$a = \dfrac{v_f - v_i}{t}$	$\alpha = \dfrac{\omega_f - \omega_i}{t}$
$v_{avg} = \dfrac{v_f + v_i}{2}$	$\omega_{avg} = \dfrac{\omega_f + \omega_i}{2}$
$a = \dfrac{2d}{t^2}$ $(v_i = 0)$	$\alpha = \dfrac{2\theta}{t^2}$ $\omega_i = 0)$

It is important to understand how the formulae are analogous, that they are conceptually the same, the only real difference being that the displacement is angular rather than translational. For example, velocity units would be revolutions per minute or radians per second rather than miles per hour or feet per second.

The rotational formulae are applied to numerical problems in much the same way as the translational formulae. In fact, the same types of quantities are involved. Like in translational problems, there is time, displacement, initial, final, and average velocity, and acceleration. The relationships between the quantities is the same. As a result, solving rotational problems is the same as translational except that the displacement is angular; things are spinning rather than moving more or less in a straight line. As a result, the problems are solved essentially the same way.

Example 2-7

Calculate the angular velocity of a flywheel that rotates 6000 revolutions in 2 minutes.
Solution:

$$\omega = \frac{\theta}{t} = \frac{6000 \text{rev}}{2 \text{min}} = 3000 \text{ rpm}$$

Example 2-8

Convert 1200 rpm to rad/s.
Solution:

Since there are 2π rads per rev and 60 sec per minute, we must multiply by 2π and divide by 60.

$$\omega = \frac{1200 \text{ rev}\left(\frac{2\pi \text{rad}}{1 \text{rev}}\right)}{1 \text{min}\left(\frac{60 \text{sec}}{1 \text{min}}\right)} = 126 \text{ rad/s} \quad \textit{(Rounding to three significant digits.)}$$

Example 2-9

A brake is applied to a rotating shaft for 4 seconds, reducing the angular velocity from 1200 rpm to 540 rpm. Find (a) the angular "acceleration" in rad/s^2 and (b) the angular displacement in radians.
Solution:

$$\omega_i = 1200 \text{ rpm} = 126 \text{ rad/s}$$
$$\omega_f = 540 \text{ rpm} = 56.5 \text{ rad/s}$$

a. $\quad \alpha = \dfrac{\omega_f - \omega_i}{t} = \dfrac{56.5 \text{rad/s} - 126 \text{rad/s}}{4s} = -17.4 \text{rad/s}^2$

b. $\quad \omega_{avg} = \dfrac{\omega_f + \omega_i}{2} = \dfrac{56.5 \text{rad/s} + 126 \text{rad/s}}{2} = 91.3 \text{ rad/s}$

$\quad \omega = \dfrac{\theta}{t}; \quad \theta = \omega t = 91.3 \text{rad/s}(4s) = 365 \text{ rad}$

Student Name(s)_____

_____

_____

Measuring Angular Rate with a Stroboscope

CAUTION! *The intense flashing of the strobe can hurt the eyes. It can also cause dizziness and disorientation. Strobe lights can induce seizures in people prone to certain disorders. So it is very important to limit the amount of time the strobe is turned on and to refrain from looking directly into the light.*

Objectives

When you've finished this lab, you should be able to do the following:
- Measure the rate of a high-speed rotating object with a stroboscope.
- Express angular rate in revolutions per minute and an equivalent radians per second

Main Ideas

- Angular rate is a measure of how fast an object rotates or spins.
- Angular rate can be measured in revolutions per minute (rpm) or radians per second.
- A stroboscope may give false readings is it's not checked at multiples of the rate shown on the dial or display.

Discussion

One can find a motor's rotational speed by measuring its angular rate. In Industry, many devices are turned by motors. These devices include pumps, conveyor belts, mixers, and compressors. Each device is designed to turn at a certain speed. It is important for technicians to know how to control and measure these angular speeds for safe and efficient operation. One common example of stroboscopes in use is the timing light used on automobile engines: triggered by the voltage at a spark plug, the rate of the firing of the plug is matched to the rate of the mark on the turning shaft.

A device called a **stroboscope** can be used to measure rotational speeds. A stroboscope is a device that emits short flashes of intense light at definite and controllable intervals. The rate of the strobe flashes can be adjusted. Usually, the display on the strobe is in flashes-per-minute. A mark is made on the rim of the rotating device. When the rate of the flashing strobe matches the rate of the rotating machine, an illusion is created: the rotating device appears to be stationary. This is because when the strobe is off, the rotating device is blurred. But when the strobe flashes on, our eyes register the mark in the same place each time it makes a revolution, making appear stationary. This means that the revolutions-per-minute of the rotating device is equal to the flashes-per-minute of the strobe.

The **problem of multiples** can cause erroneous readings. This happens because the illusion of the device being stationary can be caused by multiple turns, not just one. In other words, the illusion can be caused by two, three, or more turns of the device, not only one. This problem can be alleviated with a simple test: double the strobe rate. If the initial reading is correct, two marks should be seen, one halfway around (the flashes catch the mark twice for each revolution). If this doesn't happen, the reading is incorrect. The strobe must then be adjusted for the illusion to occur at a different setting, followed again by the doubling test.

NOTE: Because of this problem with multiples, it is often very difficult to get a first reading that is correct, a correct base reading. You may ask your instructor to use a tachometer to provide your first base reading or for general comparisons..

Angular rates are often measured in revolutions per minute (rpm), but other units of measurement are also commonly used, especially when using formulae. Here we will **convert rpm to radians** per second (rad/sec). Two mathematical operations must be performed, one to convert revolutions to radians and another to convert minutes to seconds. Since there are 2π radians in a revolution, we multiply by 2π to convert revs to rads. Since there are sixty seconds in one minute, we divide by 60 to convert from minutes to seconds. As a result, when converting from rpm to rad/sec, we must multiply by 2π then divide by 60. When converting from rad/sec to rpm, the reverse is done.

Equipment

Stroboscope
DC electric motor
DC power supply
Laser Tachometer
Optional: any rotating device such as a cooling fan

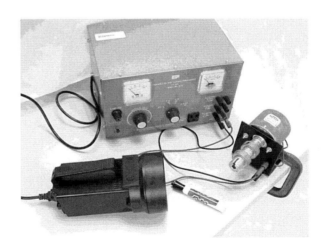

Procedure

1. Study the operation of the stroboscope. Be sure you understand how to control the rate of flashing. Most stroboscopes have two adjustment controls, one for fine adjustment and one for wider ranges. There also might be a range button, one position it is set for high speed flashing and the second position for slow. **Do not look directly into the light. Do not point the light at anyone. Shut the strobe off as often as possible.**

2. Make a clear mark on the rim of the rotating device or one the shaft of the electric motor. (Or a piece of tape stuck on the shaft can serve as a flag.)

3. Connect the motor to the DC power supply. If the power supply has a built-in meter, use this meter to set the voltage. If the power supply does not have a built-in meter, a voltmeter must be connected in the circuit.

4. Set the voltage on the DC power supply to 12 volts (high motor speed is easiest to begin with because the still-illusion does not work well at lower speeds; the eye is not so easily fooled).

5. Turn the strobe on and point it at the mark. Adjust the strobe rate until it appears still. Check for the problem of multiples as described above. Readjust as necessary. Record this value in the data table.

6. For comparison, use the tachometer to measure the speed, recording this value in the table. Apply either aluminum tape or the reflective tape included in the tachometer kit to the rotating device as a flag for the strobe. The strobe and tachometer readings should be approximately equal, with flashes per minute equal to rev/min.

7. Repeat step 5 for the remaining voltage settings, checking each reading for multiples and recording the values in the data table. As the voltage is decreased, expect a proportional decrease in the angular speed.

8. Convert the rpm readings to rad/sec, recording the conversion in the data table.

Data Table

Voltage Setting	Stroboscope Reading (flashes per minute)	Tachometer Reading (rpm)	Angular Speed (rad/sec)
12 V			
10 V			
8 V			
6 V			
4 V			

Questions

Suppose a wheel is turning at a rate of 3600 rpm (60 revolutions per second). A strobe light is used to measure the wheel's rate. It's set a rate of 1800 fpm (30 flashes per second).

1. Will the wheel appear to stand still? Explain.

2. What happens when the strobe rate is increased to 60 flashes per second? Will the wheel appear to stand still? Explain.

3. What happens at 120 flashes per second? Explain.

4. Which flashing rate of the strobe is the correct one? Explain.

1. What is the unifying equation for rate?

2. How does the unifying equation for rate apply to rotational systems?

3. In rotating mechanical systems, displacement is given as the _____ through which the body rotates.

4. Convert 2000 rpm to rad/sec.

5. The scanner dish of a weather radar antenna system completes 160 revolutions in one hour. Calculate the angular speed of the dish in rpm and rad/sec.

6. A rotating shaft "accelerates" from 1200 rpm to 500 rpm in 5 seconds. Calculate
   a. the average angular velocity in rad/sec.

   b. the angular displacement.

   c. the angular acceleration in rad/s^2.

Answers: 1. rate is the displacement divided by time **2.** The angular displacement (or angle) divided by time **3.** angle **4.** 209 rad/s **5.** 2.67 rpm, 0.279 rad/s **6.a** 89.0 rad/s **6.b** 445 rad **6.c** -14.7 rad/s^2

Fluid systems are common in residential, commercial, and manufacturing applications. Not only do fluid systems keep the water running in our homes, but fluid systems are closely related to thermal cooling systems that keep other systems running safely and efficiently.

Liquids and gases are both fluids. Both types of fluids behave in exactly the same way but for one major difference; liquids generally cannot be compressed, but the volume of gasses can easily change under changes in pressure. In a **hydraulic** system a liquid such as water or oil is used, while **pneumatic** systems use a gas such as air. Hydraulic systems might drive pistons on tractor loaders. A toilet is a hydraulic system. Pneumatic systems power many tools such as nail guns and air wrenches. Some control systems are operated pneumatically, systems that regulate and control heating and cooling systems, automated machining tools, and many other applications.

Following our unifying principle of displacement over time, **fluid flow rate** is the rate at which a volume of fluid is displaced. In other words, the fluid flow rate is the amount of fluid volume divided by the time interval in which the displacement took place.

$$\textbf{FLUID FLOW RATE} = \frac{\textbf{DISPLACED VOLUME OF FLUID}}{\textbf{ELAPSED TIME}}$$

We discovered in Chapter 1 there are many units of measurement that may be used to quantify fluid volume. These include units of length cubed such as ft^3 and m^3 to fluid measures such as gallons and liters. The amount of fluid can also be measured by its mass, in grams and slugs, or its weight in pounds and Newtons. (More on mass verses weight later.) Any of these measures divided by any units of time are legitimate units of fluid flow rate, but of course there are certain conventional standards of measurement, gallons per minute (gpm), cubic feet per hour (ft^3/h) or (cfh), cubic feet per second (ft^3/s), cubic meters per second (m^3/s), and kilograms per second (kg/s).

There are many, many different types of rate because there are many different types of displacement. As a result, generating new literals to represent all these rates in formulae can be confusing and cumbersome. So a system or writing was designed that makes it easy to write symbols for all these different rates. This writing/naming system is not universal, but is commonly used in technical literature. It is now time to introduce this particular system of writing rate formulae.

The displacement quantity for fluid flow rate is the volume, represented by capital letter "V". A symbol to represent volume flow rate is created by using the letter for the displacement, V, and writing a dot over it, $\dot{V}$. Similarly, if the fluid displacement quantity is the mass (m), mass flow rate can be symbolized with $\dot{m}$. If the displacement quantity is the weight (w) of a fluid,

fluid weight flow rate can be symbolized with $\dot{w}$. So whatever the displacement letter is, just put a dot over it and you have a symbol representing its flow rate when different types of displacement quantities are used.

$$\text{Volume Flow Rate} \quad \dot{V} = \frac{V}{t}$$

$$\text{Mass Flow Rate} \quad \dot{m} = \frac{m}{t}$$

$$\text{Fluid Weight Flow Rate} \quad \dot{w} = \frac{w}{t}$$

In sections 2.1 and 2.2 objects were accelerated, with velocities changing within time intervals. Although fluid flow rates do indeed accelerate, we will not consider fluid flow rate accelerations here, limiting discussion to straightforward, un-accelerated fluid rates.

Example 2-10

It takes 2.5 hours to fill a 1000 gallon swimming pool. Calculate the fluid flow rate of the household water pump in gpm.
Solution:

$$t = 2.5 \text{ hr} \left(\frac{60\text{min}}{1\text{ hr}}\right) = 150 \text{ min}$$

$$\dot{V} = \frac{V}{t} = \frac{1000\text{gal}}{150 \text{ min}} = 6.67 \text{ gpm}$$

Example 2-11

Five million gallons of fuel is being pumped off a ship by two 850 gpm pumps working in tandem. How many hours will it take to completely unload the ship?
Solution:

$$\dot{V}_T = 2(850\text{gpm}) = 1700\text{gpm}$$

$$\dot{V} = \frac{V}{t} \; ; \quad t = \frac{V}{\dot{V}} = \frac{5 \times 10^6 \text{gal}}{1700\text{gpm}} = 2.94 \times 10^3 \text{ min} = 49 \text{ hours (or about two days)}$$

(The displacement is always divided by the rate to calculate the time. Just make sure the units are consistent.)

Example 2-12

200 kg of ammonia gas flows through the cooling coils of a refrigeration unit in 5 minutes. Calculate the mass flow rate of the gas.
Solution:

$$\dot{m} = \frac{m}{t} = \frac{200\text{kg}}{5\text{min}} = 40 \text{ kg/min}$$

1. What is the unifying equation for rate?

2. How does the unifying equation for rate apply to fluid systems?

3. Six million cubic feet of water flows over the crest line of Niagara Falls each minute. Find the flow rate in gpm. Leave answer in scientific notation.

4. A pump provides 12,000 gallons of water in ten hours. Calculate the fluid flow rate in gpm.

5. A large fuel truck delivers fuel to a large municipal building with a 3,000 gallon fuel tank. The pump on the fuel truck pumps the fuel at a rate of about fifteen gallons per minute. The fuel truck driver is not allowed to leave the pump nozzle in the fuel filler pipe unattended. How long must the man stand outside in a twenty below zero wind chill to fill the building's tank?

6. For health reasons, air must be circulated, or exchanged, in a room every six to ten minutes. An air circulation system with a mass-flow rate of 14.6 kg/min circulates the air in a room of size 6m x 6m x 2.5m. This room holds 117 kg of air. Does the system satisfy the design requirement? Support your answer with calculations.

7. How long will it take to fill a 20,000 gallon tank at a rate of 5 gpm?

8. A supertanker carries 84.0×10^6 gallons of crude oil. The oil must be unloaded at an offshore buoy within five days. Calculate the fluid flow rate required.

Answers: 1. rate is the displacement divided by time **2.** Fluid rate is the fluid displaced divided by time **3.** 4.50×10^7 gpm **4.** 20 gpm **5.** 3.33 hrs **6.** yes, 8.00 min **7.** 66.7 hrs **8.** 700,000 gph

In the Chapter 1 section on displacement in electrical systems, we learned that the displacement quantity in electrical systems is the charge (q), measured in coulombs (c), an amount of electrons. Following our unifying principle of displacement over time, and borrowing a term from analogous fluid systems, electric **current** is the rate at which charge is displaced.

Electrical current is the "speed" of the electrons flowing through conductors.

$$\textbf{ELECTRIC CURRENT} = \frac{\textbf{DISPLACED CHARGE}}{\textbf{ELAPSED TIME}}$$

Current is represented by the capital letter "I" in formulae. When units of charge are divided by units of time, the result in coulombs per second or c/s. The coulomb per second was named after the scientist André-Marie Ampère, often simplified to "amp", and abbreviated as capital "A".

Electric Current: $I = \frac{q}{t}$

Units: 1A = 1 c/s

André-Marie Ampère (1775 – 1836)

André-Marie Ampère was a French physicist and mathematician who is generally regarded as one of the main founders of the science of classical electromagnetism, which he referred to as "electrodynamics". The SI unit of measurement of electric current, the ampere, is named after him.

Example 2-13

The label on a 12 volt battery says the battery contains 350,000 coulombs. The battery provides a steady current to an electrical load for 6 hours until the battery finally dies. Calculate the average current provided by the battery during this time interval.

Solution:

$$t = 6 \text{ hours} = 21,600 \text{ seconds}$$

$$I = \frac{q}{t} = \frac{350,000 \text{ c}}{21,600 \text{ s}} = 16.2 \text{ A}$$

2.4 STUDENT EXERCISES   ELECTRICAL RATE

1. The displacement in electrical systems is the amount of _____.

2. In electrical systems, the rate at which charge is displaced is the _____.

3. How does the unifying equation for rate apply to electrical systems?

4. Calculate the average current when 2.81×10^{24} electrons are displaced by a battery over a period of ten hours.

5. A 12 volt battery is said to contain 200,000 coulombs. Calculate how long this battery should last when delivering a steady ten amperes.

6. Explain the difference between electrical charge and electrical current.

Answers: **1.** charge **2.** current **3.** electrical rate is the current, which is the charge divided by the time **4.** 12.5A **5.** 5.56 hrs **6.** charge is the measure of the amount of electrons available (or displaced) while current is the rate at which charge moves through a conductor

From Chapter 1 we learned that heat (Q) is the displacement quantity in thermal systems. We learned about sensible and latent heat and that the relationship between the temperature and heat varies from substance to substance, a property called specific heat. When a temperature difference exists, heat will move from a higher temperature region to a lower temperature region. Following our unifying principle for rate, heat flow rate is the rate at which heat flows from the warmer region to the colder, or heat divided by time.

$$\text{HEAT FLOW RATE} = \frac{\text{DISPLACED HEAT}}{\text{ELAPSED TIME}}$$

The symbol used in formulae to represent heat flow rate is written much like the fluid flow rate dotted V. The displacement literal for heat, Q, is dotted to represent heat flow rate, $\dot{Q}$.

Heat Flow Rate

$$\dot{Q} = \frac{Q}{t}$$

The units of measurement for heat flow rate are fairly straightforward. In the English system the standard is Btu/hr or Btu/min. In SI the units are cal/sec or kcal/sec.

Example 2-14

A space heater rated 7500 Btu/hr runs continuously for 4 hours. Calculate the amount of heat energy produced.

Solution:

$$\dot{Q} = \frac{Q}{t}; \quad Q = \dot{Q}t = 7500\,\text{Btu/\sout{hr}}(4\,\text{\sout{hrs}}) = 30{,}000\ \text{Btu}$$

Specifications on equipment labels and plates often leave off the "per hour", writing Btu but really meaning Btu/hr. Here the "per hour" will not be omitted, but technicians must be conscious of this omission of time units.

Example 2-15

A chef's hot plate is rated 800 cal/sec. How much time will it take for this hot plate to completely boil away 1 kg of water that is initially at room temperature, 20°C?

Solution:

$m = 1\text{kg} = 1000\text{g}$

$\Delta T = 100 - 20 = 80\ \text{C}°$

$$Q_T = mc\Delta T + mH_v = 1000\text{g}\left(1\,\tfrac{\text{cal}}{\text{g·C}°}\right)(80\,\text{\sout{C}°}) + 1000\text{g}\left(540\,\tfrac{\text{cal}}{\text{g}}\right) = 620{,}000\text{cal}$$

$$t = \frac{Q}{\dot{Q}} = \frac{620{,}000\text{cal}}{800\text{cal/sec}} = 775\ \text{sec} = 12.9\ \text{min}$$

2.5 STUDENT EXERCISES   THERMAL RATE

1. What is the unifying equation for rate?

2. How does the unifying equation for rate apply to thermal systems?

3. A lumber company uses a drying oven that produces 300,000 Btu during a half hour drying cycle. Calculate the heat flow rate of the oven in Btu/hr.

4. A sterilizer heating element used in a medical facility is rated 1000 cal/min. 1.8×10^5 calories is needed to properly sterilize instruments. Calculate the amount of time the heating element must operate to meet the standard.

5. A refrigeration unit on a delivery truck is rated 40,000 Btu/hr. How much heat will the unit remove if it operates for 3.5 hours during an 8 hour on/off cycle?

6. What is the heat flow rate that will heat 45 gallons of water from room temperature (70°F) to 160 °F in 15 minutes?

Answers: 1. in general, rate is the displacement divided by the time 2. thermal rate is the heat displaced divided by the time 3. 600,000 Btu 4. 3 hrs 5. 140,000 Btu 6. 135,000 Btu/hr (rounding to three significant digits, as always for non-lab calculations)

Chapter 3   The Movers

We have just completed two chapters, one on displacement quantities and another on the rate at which these displacement quantities are moved. But what causes all this movement, this displacement? This chapter addresses the quantities that *cause* this movement, what we will call the "Movers".

In each energy system there is a quantity that causes movement or some kind of change. These quantities are the movers. The **movers** cause displacement.

The Unifying Principle:

Each energy system has a quantity that causes movement or change. We will call these quantities the "MOVERS"

The Movers in Each Energy System

Translational Mechanical: **Force** (F) moves objects.

Rotational Mechanical: **Torque** (τ) turns objects.

Fluid: **Pressure** (p) moves fluid.

Electrical: **Voltage** (E or V) moves charge.

Thermal: **Temperature Difference** (ΔT) moves heat.

A **force** is a push or a pull that causes objects to move or to stop moving. Force is the "mover" in translational mechanical systems.

Most people are familiar with the units of force in the English system, the **pound**, abbreviated "lb". In the metric system, the units of force is the Newton, abbreviated "N", named after Isaac **Newton**, who formalized the distinction between mass and weight. Mass verses weight is discussed later in this section.

Force is a vector quantity with magnitude and direction. In diagrams, forces are drawn as arrows. Much like the displacement vectors we studied in Chapter 1, the length of the vector arrow represents the magnitude and the direction the arrow points represents the direction of the force.

Sometimes many different forces interact. Trusses, frames, and other structures are designed to support loads in a variety of conditions. In these situations force vectors must be added using trigonometric methods similar to what we did with the displacement vectors. A **resultant force** is the vector sum of two or more force vectors. Forces are added by drawing their vectors **tip-to-tail** to produce a triangle. It is from this triangle forces can be added using trigonometric techniques. Subscripts are often used to distinguish between different forces. For example, two forces adding up to a total, or resultant, force might be written $F_T = F_1 + F_2$. Here we will limit discussion to force vectors that can either be added directly or are interacting perpendicular to each other and thus forming right triangles.

It's easy to add vectors when they're on the same **line of action**; the resultant is a simple sum. (Be careful about what is meant here by a "sum". A sum here is an algebraic sum, where the answer could be the result of subtraction. Algebraically, we're always "adding" signed numbers, both positive and negative.)

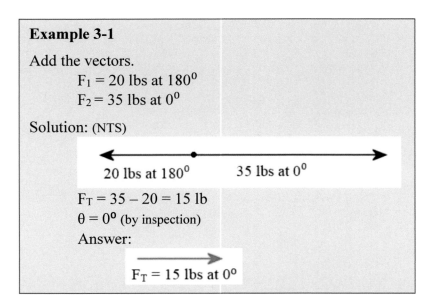

Example 3-1

Add the vectors.

$F_1 = 20$ lbs at 180^0
$F_2 = 35$ lbs at 0^0

Solution: (NTS)

20 lbs at 180^0 35 lbs at 0^0

$F_T = 35 - 20 = 15$ lb
$\theta = 0^0$ (by inspection)
Answer:

$F_T = 15$ lbs at 0^0

The force vectors can be "added" directly when they are on the same line-of-action. The direction of the resultant vector is along this line of action, but in the direction of the larger vector. We can begin any complex vector addition problem by first adding these line-of-action vectors as a good start to combining them, adding what's left by trigonometry later.

When forces do not act on the same line of action, adding them gets a little more complicated because we can't just add them directly. When the forces interact at different angles, trigonometry must be used to calculate the resultant. We will limit discussion here to right triangles.

Example 3-2

Add the forces.

$F_1 = 3$ lbs at 0^0

$F_2 = 4$ lbs at 90^0

Solution: (NTS)

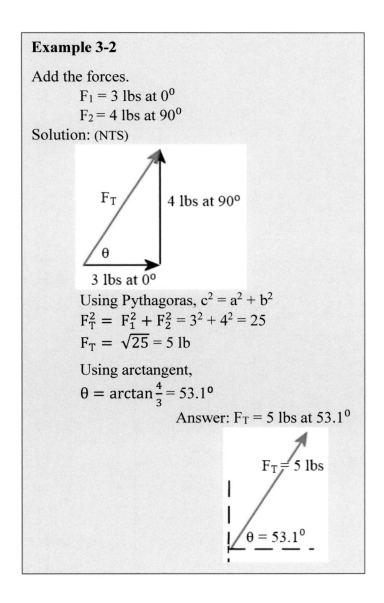

Using Pythagoras, $c^2 = a^2 + b^2$

$F_T^2 = F_1^2 + F_2^2 = 3^2 + 4^2 = 25$

$F_T = \sqrt{25} = 5$ lb

Using arctangent,

$\theta = \arctan\frac{4}{3} = 53.1^0$

Answer: $F_T = 5$ lbs at 53.1^0

Example 3-3

Calculate the resultant vector.

$F_1 = 75$ kN at 180^0
$F_2 = 90$ kN at 90^0
$F_3 = 110$ kN at 270^0

Solution: (NTS)
We can begin by directly adding the two vectors that are on the same vertical line of action.

$F_2 + F_3 = 90 - 110 = 20$ kN at 270^0

Three given vectors:

$F_2 = 90$ kN at 90^0

$F_1 = 75$ kN at 180^0

$F_3 = 110$ kN at 270^0

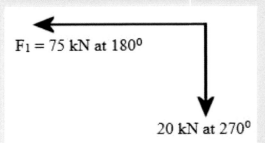

$F_1 = 75$ kN at 180^0

20 kN at 270^0

The remaining two vectors form a right triangle. The vectors are redrawn tip-to-tail to form a triangle.

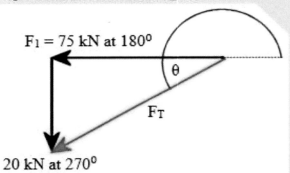

$F_1 = 75$ kN at 180^0

θ

F_T

20 kN at 270^0

$F_T^2 = 75^2 + 20^2 = 6025$

$F_T = \sqrt{6025} = 77.6$

$\theta = 180 + \arctan \frac{20}{75} = 194.9^0$

Answer: $F_T = 77.6$ kN at 194.9^0

126

Resolving Vectors

In the above problems we were given forces that were acting either vertically or horizontally that were working together to produce one resultant force. We then calculated their resultant, their vector sum. Now we investigate the reverse situation: we are given a vector, one that might have a direction at any angle, and we must determine how much of it is acting purely in the vertical direction and how much of it is acting in the purely horizontal direction. This is sometimes called "finding the x and y components of a vector", or simply "resolving" the vector. **Resolving the vector** is calculating its x and y components. In this situation the magnitude and direction of a single vector are given and the remaining opposite and adjacent sides are determined using trigonometry.

Recall that $\sin \theta = \frac{opp}{hyp}$. Given the angle ($\theta$) and the hypotenuse, the opposite side can be calculated using opp = hyp (sin θ).

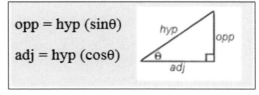

Recall that $\cos \theta = \frac{adj}{hyp}$. Given the angle ($\theta$) and the hypotenuse, the adjacent side can be calculated using adj = hyp (cos θ).

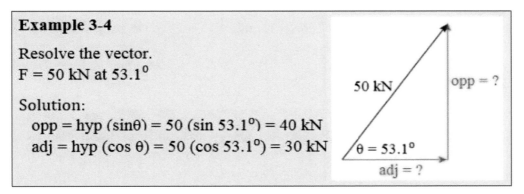

Example 3-4

Resolve the vector.
F = 50 kN at 53.1°

Solution:
  opp = hyp (sinθ) = 50 (sin 53.1°) = 40 kN
  adj = hyp (cos θ) = 50 (cos 53.1°) = 30 kN

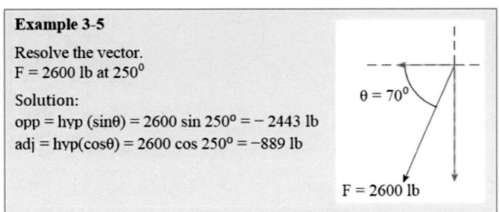

Example 3-5

Resolve the vector.
F = 2600 lb at 250°

Solution:
opp = hyp (sinθ) = 2600 sin 250° = – 2443 lb
adj = hyp(cosθ) = 2600 cos 250° = –889 lb

Newton's Law

We learned about acceleration in Chapter 2. Now we're ready to learn about one of Isaac Newton's greatest, yet simplest, scientific achievements: the mathematical relationship between mass, force, and acceleration. This is known as Newton's Law, the foundation of all materialistic science. This law explains the motions of all objects, including cars, planets, rockets, and you. **Newton's Law** states that an object's acceleration is directly proportional to the force acting on it but inversely proportional to its mass.

The more massive an object is, the more difficult it is to get moving or to slow down when it's already moving, so this quantity is in the denominator (inversely proportional). An increase in force will of course increase the acceleration.

Mass is a measure of inertia. **Inertia** is the tendency of mass to stay still if still, but if moving, to tend to keep moving until acted on by force.

When units of mass are multiplied to units of acceleration in F = ma, we get kg·m/s^2 in the metric system. It is this cumbersome combination of units that was named after Newton.

> Newton's Law
>
> $$a = \frac{F}{m}$$
>
> or
>
> **F = ma**

$$1 \text{ N} = 1 \text{ kg·m/s}^2$$

When units of mass are multiplied to units of acceleration in F = ma, we get slug·ft/s^2 in the English system. This combination of units is the pound.

$$1 \text{ lb} = 1 \text{ slug·ft/s}^2$$

Example 3-6

How much force is needed to accelerate a 60 kg object to a speed of 20 m/s in 5 seconds?

Solution:

$$a = \frac{\Delta v}{t} = \frac{20\frac{m}{s} - 0\frac{m}{s}}{5s} = 4 \text{ m/s}^2$$
$$F = ma = 60kg(4m/s^2) = 240 \text{ N}$$

Mass Verses Weight

There's a reason why the unit of force in the metric system is named after this iconic scientist. Isaac Newton is given credit for being the first to mathematically clarify the distinction between mass (m) and weight (w). He was the first to realize there is something more fundamental than the weight of an object, something more basic, its **mass.**

Mass is the total amount of matter of an object. Mass is measured in slugs (slugs) in the English system and grams (g) in the metric system. The object's size, or volume, is irrelevant. The amount of air in a large room could be the same mass as a small piece of lead. Also, exactly what the substance is also is irrelevant, whether it be Styrofoam, lead, air, or gold; mass is just the amount of material. Mass is a scalar quantity having only magnitude.

Weight, on the other hand, is the force created when gravity acts on mass. Weight is a vector quantity having both magnitude and direction. The direction is always downward, or 270°. As mentioned earlier, the units of weight are units of force, the pound (lbs) in the English system and the Newton (N) in the metric system.

What sets mass and weight apart is gravity. The weight of an object can change with gravity. An object on the moon has one sixth the weight it would have on Earth. Sometimes objects can be considered weightless. But the amount of mass remains constant regardless of the weight.

Weight and Newton's Law

Newton's Law, $F = ma$, is the general formula that applies to all objects experiencing any acceleration by any force. But this formula also applies to the weight of an object, where the acceleration is constant, the acceleration due to gravity. We learned about acceleration due to gravity back in our chapter on rate.

Weight is a type of force. When Newton's Law is applied to weight, the acceleration is the acceleration due to gravity, a constant. The formula for mass vs weight is essentially the same as $F = ma$ except that instead of "F" for force, "w" for weight is used. And instead of "a" for acceleration, "g" for acceleration due to gravity is used, a particular and constant value of acceleration.

Acceleration Due To Gravity
English System: $g = 32.2$ ft/s^2
Metric System: $g = 9.81$ m/s^2

Newton's Law Applied to Weight
$w = mg$

Example 3-7

How much does a 10 kg object weigh?
Solution:
$$w = mg = 10 \text{ kg } (9.81 \text{ m/s}^2) = 98.1 \text{ N}$$

Example 3-8

What is the mass of a 12 pound weight?
Solution:
$$m = \frac{w}{g} = \frac{12 \text{ lb}}{32.2 \frac{\text{ft}}{\text{s}^2}} = 0.373 \text{ slugs}$$

Sir Isaac Newton (1642 – 1727)

Sir Isaac Newton was an English physicist and
mathematician (described in his own day as a
"natural philosopher") who is widely recognised
as one of the most influential scientists of all time
and as a key figure in the scientific revolution.
His book *Philosophiæ Naturalis Principia Mathematica*
("Mathematical Principles of Natural Philosophy"),
first published in 1687, laid the foundations for classical
mechanics. Newton also made seminal contributions to
optics and shares credit with Gottfried Leibniz for the
invention of calculus.

Newton's *Principia* formulated the laws of motion and universal gravitation, which
dominated scientists' view of the physical universe for the next three centuries. By
deriving Kepler's laws of planetary motion from his mathematical description of gravity,
and then using the same principles to account for the trajectories of comets, the tides, the
precession of the equinoxes, and other phenomena, Newton removed the last doubts
about the validity of the heliocentric model of the cosmos. This work also demonstrated
that the motion of objects on Earth and of celestial bodies could be described by the
same principles. His prediction that the Earth should be shaped as an oblate spheroid was
later vindicated by the measurements of Maupertuis, La Condamine, and others, which
helped convince most Continental European scientists of the superiority of Newtonian
mechanics over the earlier system of Descartes.

Student Name(s) _____

_____

_____

_____

OBJECTIVES:

Upon completion of this lab, the student will be able to

1. discriminate between vectors and scalars.
2. describe vectors and vector components as related to force.
3. draw scale vector diagrams.
4. calculate the components of a vector using trigonometry.
5. define and calculate net force.
6. define net force as related to a mechanical system in equilibrium.
7. experimentally verify the method of vector addition by components.

DISCUSSION:

A great many quantities are fully described only by their magnitudes (how much). These are the *scalar quantities* include mass, which might be measured in grams, volume in cubic feet, time in seconds, and heat in BTUs.

Many other quantities are fully described only when both their magnitude *and direction* (their angles) are specified. Such quantities are known as *vectors*. Examples of vectors are velocity, momentum, voltage (often referred to as a phasor), and of course, force, which is studied here. Vectors are of utmost importance in many fields of science and technology, with applications ranging from three phase voltage systems to roof trusses to automobile frame straightening machines.

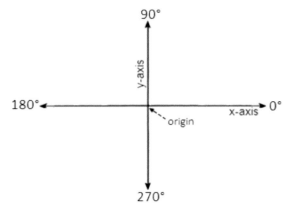

Vectors are represented graphically using arrows. The length of the arrow represents its magnitude (how much) and the angle at which the arrow points represents its direction. Vectors are drawn on the standard x-y axis grid with zero degrees located horizontally at right then rotationally increasing counter-clockwise.

When a vector is drawn on this x-y grid, its angle is first marked using a protractor. Then a scale is established, say one centimeter represents ten pounds. We draw the vector by beginning the vector's "tail" at the origin then passing through the marked angle to a length determined by the pre-established scale. It is then topped off with an arrow head.

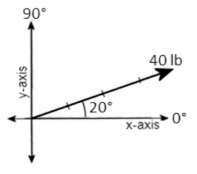

Shown here is the example of a diagram of a force vector with a magnitude of 40 pounds acting at a direction of 20°. Note how the vector's magnitude and direction are always labeled and how sometimes only the relevant grid quadrant is depicted.

When more than one force is acting within a system or upon and object, the forces combine into a *vector sum*. This sum is sometimes called the *resultant* or *net force*.

Calculating the net force is easy *when all the forces act on the same line of action*, like in the tug-of-war below. We simply "add" them up directly, but being careful to give the left acting vectors a negative sign and the right-acting vectors a positive sign. Some of the vectors are working together. Some are not. Here it is readily apparent the team on the left is winning. The net force is $-50 - 50 - 50 + 50 + 50 = -50$ lb or 50 lb to the left. Note how the answer to a vector problem always includes both magnitude and direction to fully describe the resultant vector.

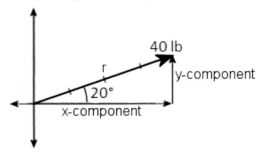

$$F_{net} = -50 - 50 - 50 + 50 + 50 = -50 lb \; or \; 50 lb \; @180°$$

Calculating net force where vectors are *not on the same line of action* is a bit more involved. ***Vector addition by method of components*** tends to be the most useful approach and requires right triangle trigonometry. The vectors in the above tug-of-war all point either directly to the left or to the right so can be algebraically added directly. They all are acting directly along the x-axis. But any other vector that does not lie directly on the x or y axis acts a bit on both axes, that is, it is acting partly horizontally and also acting partly vertically. The horizontal part of the vector is called the *x-component* and the vertical part of the vector is the *y-component*. Perpendicular, these two components form a right triangle, with the total magnitude of the vector in the hypotenuse position. Calculating how much a vector is acting on each axis is called *resolving*. An example can best illustrate this.

Earlier we drew a 40 pound vector acting at twenty degrees. Let's resolve this vector into its x and y-components.

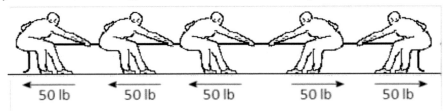

The vector has magnitude of 40 lb and angle of 20°. The two components form the opposite and adjacent sides of a right triangle with the vector magnitude in the hypotenuse position. We name the opposite side *y*, the adjacent side *x*, and the hypotenuse *r* (for resultant of the two components).

Manipulating $\sin\theta = \frac{y}{r}$ for y we have $y = r\sin\theta$. The y-component is found by multiplying the sine of the angle to the total magnitude.

Manipulating $\cos\theta = \frac{x}{r}$ for x we have $x = r\cos\theta$. The x-component is found by multiplying the cosine of the angle to the total magnitude.

The x and y-components of our example above with r = 40 lb @ 20° are then

$$y = r\sin\theta = 40 \sin 20° = 40(0.342) = 13.7\ lb$$

$$x = r\cos\theta = 40\cos20° = 40(0.940) = 37.6\ lb$$

Checking with the Pythagorean Theorem:
$$r^2 = y^2 + x^2$$
$$r^2 = 13.7^2 + 37.6^2$$
$$r^2 = 1601$$
$$r = 40.0\quad \sqrt{}$$

Imagine a man pulling upward and horizontally on a box with a 40 lb force at 20° as shown. According to the above calculations, he is pulling upward 13.7 lb and horizontally 37.6 lb. So, only 37.6 lb of his effort actually works on dragging the box. The vertical 13.7 lb is wasted (unless this force reduces friction, but that's another story).

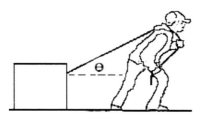

In this lab exercise three vectors act to cancel each other out. When all the forces cancel themselves out, the net force is zero, and no movement will occur. This condition is called **equilibrium**. The vectors in this activity will be arranged as shown, with two "upward" vectors cancelling each other out horizontally. These also act together to cancel the "downward" vector. Here the sum of the y-components of the two "upward" vectors should be equal and opposite the downward vector. In other words, *for equilibrium, the y-components must add up to zero and the x-components must add up to zero.*

Since the x-components are on the same line of action, they can be added directly, but being careful of sign.

Since the y-components are on the same line of action, they can be added directly, but being careful of sign.

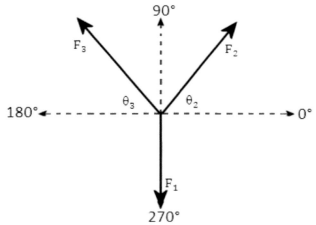

A common mistake made by students is confusing the physical *length of a string* from which a weight is suspended and the *force in the string*. **When drawing vectors, the length of the vector represents the force, not the string length**. A very short string could have a very large force and visa-versa.

EQUIPMENT: Kennon Force Table, assorted slotted weights with three hangers, protractor, rule, and scientific calculator.

PROCEDURE:

1. Adjust the force table stand so that it is reasonably level.
2. Suspend a system of three weight hangers from the force table.
3. Make sure one of the weights hangs directly at 270°, the "downward" force in the above diagram.
4. Hang two more "upward" weights in quadrants I and II (the first between 0° and 90°, the second between 90° and 180°).
5. Balance the system so that the center ring does not touch the center post. Do so by adding or removing mass from the hangers and/or adjusting the "upward" angles. (Don't be afraid to pile the weights on.)
6. The magnitude of each vector is equal to the tension in that string caused by the weight. The weight is calculated by multiplying the mass in kg to 9.81 m/s, or
   $w = mg$. Enter these magnitudes in the table provided.
7. The angle of the vector is read from the force table protractor on the force table surface. Enter these angles in the appropriate spaces in the table provided.
8. On a separate sheet, and using a straight edge, draw a clear diagram of the concurrent system of vectors as seen from the top of the force table in standard position, labeling them appropriately.
9. Calculate the y-components of the two "upward" vectors". (The "downward" vector is a pure, negative y.)
10. Calculate the net force on the y-axis.
11. Calculate the x-components of the two "upward" vectors.
12. Calculate the net force on the x-axis.

DATA TABLE:

Vector #1   Magnitude F_1_____N   Direction $\theta_1 =$ 270   degrees

Vector #2   Magnitude F_2_____N   Direction $\theta_2 =$_____ degrees

Vector #3   Magnitude F_3_____N   Direction $\theta_3 =$_____ degrees

CALCULATIONS:

$y_1 = F_1 sin\theta_1 = ($ $)(-1.00) =$ _____

$y_2 = F_2 sin\theta_2 = ($ $)($ $) =$ _____

$y_3 = F_3 sin\theta_3 = ($ $)($ $) =$ _____

Net force on y-axis = _____

$x_1 = F_1 cos\theta_1 = ($ $)(0.00) =$ 0 N

$x_2 = F_2 cos\theta_2 = ($ $)($ $) =$ _____

$x_3 = F_3 cos\theta_3 = ($ $)($ $) =$ _____

Net force on x-axis = _____

WRAP-UP:

1. If the system is in equilibrium the x-components should add up to zero. Within your available precision and accuracy, do your results support this theory? If not, why?

2. If the system is in equilibrium, the y-components should add up to zero. Within your available precision and accuracy, do your results support this theory? If not, why?

3. Define equilibrium as it relates to multiple mechanical forces acting on an object.

4. Why must mechanical forces be treated as vectors rather than scalars?

5. Add up all three forces $F_1 + F_2 + F_3$ directly without trig functions. Do these add up to zero as expected in a system in equilibrium?

Draw vector diagrams for problems 1 through 9.

1. Add the vectors: 140 lb @ 90° and 60 lb @ 270°.

2. Add the vectors: 1300N @ 30° and 2300N @ 210°.

3. Add the vectors: 65 kN @ 0° and 25 kN @ 90°.

4. Add the vectors: 65 kN @ 0° and 25 kN @ 270°.

5. Add the vectors: 25 kips @ 180° and 25 kips @ 270°.

6. Resolve the vector into its x and y components: 1000 lb @ 22°.

7. Resolve the vector: 1000 lb @ 338°.

8. Resolve the vector: 5200 kN @ 245°.

9. How much of the man's applied force acts on the box to move the box horizontally?

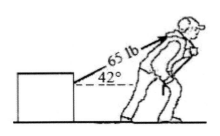

10. What is the acceleration of a 6 kg object when a force of 200 N is applied to the object horizontally?

11. How much force is needed to accelerate a 120 kg object from 10m/s to 30 m/s in 10 seconds?

12. How much does 20 slugs weigh?

13. How much does 10 kg weigh?

14. What is the mass of a 250 pounds?

15. In both English and SI units, what is the weight of a 117 kg Olympic weightlifter?

Torque is the mover quantity in rotational systems. Flywheels, gears, pulleys, fans, bolts, and screws all rotate. **Torque** is the turning or twisting action that causes this rotation. Torque results when a force is applied at some distance from an axis of rotation. This distance is called the **lever arm** or **moment arm**, symbolized by lower case ℓ, and often italicized or written long hand to differentiate the letter from the number one. "Moment" and "torque" mean essentially the same thing.

So torque depends on two quantities. Torque is the product of a force and a lever arm. The lever arm is always measured from the location of the force applied to the location of the axis of rotation.

Torque
$\tau = F\ell$

When force and distance are multiplied, the resulting units are ft-lb or in-lb in the English system and N-m in the metric system.

Example 3-9

Calculate the torque created by a 4000 pound force with a 2 foot lever arm.

Solution:

$\tau = F\ell = 4000\text{lb}(2\text{ft}) = 8000 \text{ ft-lb} = 8 \text{ kip-ft}$

Note how in example 3-9, "kilo" is at times invoked in the English system where 1000 pounds is equal to 1 kilopound and abbreviated 1 "kip". This convention is commonly used for torque (moment) measurements in structures science.

Example 3-10

Calculate the force that can create torque of 45 N-m with a 20 cm lever arm.

Solution:

$\ell = 20 \text{ cm} = 0.20 \text{ m}$

$F = \tau/\ell = 45 \text{ N-m}/0.20 \text{ m} = 225\text{N}$

Direction of Torque

Although torque isn't typically considered a vector quantity, torque does indeed have direction, a rotational direction. A plus or minus sign convention has been established to define the **direction of torque**. Recall the location of 0°, 90°, 180°, and 270° on the standard x,y coordinate system. Counterclockwise rotation is positive. Clockwise rotation is negative. (Understand this has nothing to do with clocks, it's just that clocks are a familiar way to communicate direction of rotation.)

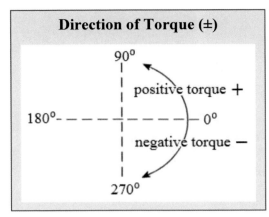

Note that the direction of the force (up, down, left, or right) is not entirely how direction of torque is determined. It's the direction of the twisting action, not the direction of the force alone that determines direction of rotation. The next few examples illustrate how a positive or negative sign is attached to the torque. A signed number is not substituted into the torque formula; the sign designation for torque is done by inspection and is attached to the answer after the torque calculation.

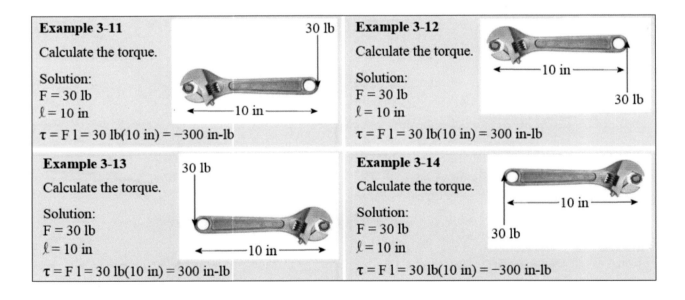

Example 3-11

Calculate the torque.

Solution:
F = 30 lb
ℓ = 10 in

τ = F l = 30 lb(10 in) = −300 in-lb

Example 3-12

Calculate the torque.

Solution:
F = 30 lb
ℓ = 10 in

τ = F l = 30 lb(10 in) = 300 in-lb

Example 3-13

Calculate the torque.

Solution:
F = 30 lb
ℓ = 10 in

τ = F l = 30 lb(10 in) = 300 in-lb

Example 3-14

Calculate the torque.

Solution:
F = 30 lb
ℓ = 10 in

τ = F l = 30 lb(10 in) = −300 in-lb

Perpendicular Lever Arm

One very important thing to know about the torque formula $\tau = F\ell$ is that the force and the lever arm must be perpendicular, that is, intersecting at ninety degrees. If the force does not act perpendicular to the lever arm, then we must calculate how much of this force does. This is usually done by resolving the force vector as shown in section 3.1.

Example 3-15

Calculate the torque.

Solution:
The vertical component of the applied force is perpendicular to the lever arm.
$F = 30\sin56^{o} = 24.9$ lb
$\tau = F\ell = 24.9$ lb(10 in) $= -249$ in-lb

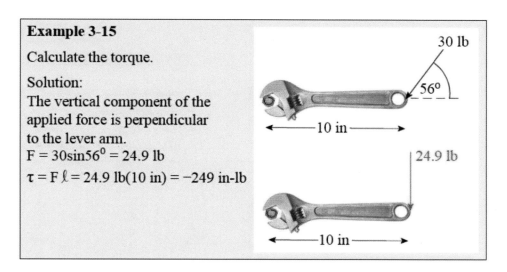

In example 3-15, only the vertical y-component of the given 30 pound force is perpendicular to the lever arm. The horizontal component of the force points directly at the nut, the axis of rotation. Any force pointing directly at, or away from, the axis of rotation does not cause torque because its lever arm is zero. Trigonometry must be employed to resolve the vector and calculate the amount of force perpendicular to the lever arm.

Example 3-16

Calculate the torque.

Solution:
The force and lever arm appear to be perpendicular. No other angle is given.
$\tau = F\ell = 54$N(25cm) $= -1350$ N-cm

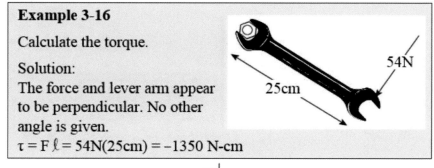

In example 3-16 the entire system is tilted or rotated, but this doesn't matter; only the force, lever arm, and their intersecting angle is of concern when calculating torque. Also, in this problem no angle between the force and lever arm is indicated.

Sometimes a box is drawn ⌐⎦ at their point of intersection to indicate 90°, but none is shown. Here the force and lever arm *appear* to be perpendicular, or near perpendicular, in the given diagram. Given no other information on the interesting angle, we must assume the angle is 90° to solve the problem. This reasonable assumption is stated in the solution.

Resultant Torque

More than one torque can be acting on a system. The total amount of torque on the system is called the "net" torque or "resultant" torque. The **net torque** is the sum of all the different values of torque. Each torque must be calculated individually first, then, being careful of sign, these are "added" together for the net, or total, torque.

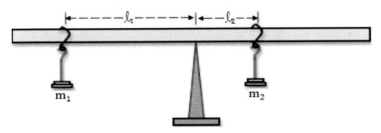

Net Torque

$$\tau_{net} = \tau_1 + \tau_2 + \tau_3 + \cdots$$

There can be any number of torques acting on a system, but in the system shown at right, two opposing values of torque "add" for a net torque. If the torque values are different, the system will rotate in the angular direction of the larger torque. But if the torques are equal, the system will balance, and will not rotate. When the mover quantities in a system completely cancel out and no displacement or motion results, the system is said to be in **equilibrium**. In this case, **rotational equilibrium** occurs when the net torque is zero and there is no rotation.

In calculating any torque problem there are two parts to the answer: the magnitude of the torque and also its angular direction. The angular direction can be expressed with clockwise (cw) or counterclockwise (ccw), but in physics the angular direction is typically expressed mathematically with a plus or minus sign. To avoid the confusion of double negatives (say −40 N-cm clockwise), these two ways should not be used simultaneously, but one or the other.

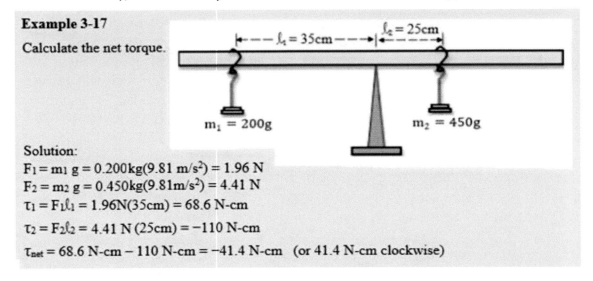

Example 3-17

Calculate the net torque.

Solution:

$F_1 = m_1\, g = 0.200\text{kg}(9.81\ \text{m/s}^2) = 1.96\ \text{N}$

$F_2 = m_2\, g = 0.450\text{kg}(9.81\text{m/s}^2) = 4.41\ \text{N}$

$\tau_1 = F_1\ell_1 = 1.96\text{N}(35\text{cm}) = 68.6\ \text{N-cm}$

$\tau_2 = F_2\ell_2 = 4.41\ \text{N}\,(25\text{cm}) = -110\ \text{N-cm}$

$\tau_{net} = 68.6\ \text{N-cm} - 110\ \text{N-cm} = -41.4\ \text{N-cm}$ (or 41.4 N-cm clockwise)

Variations in Answers Due to Rounding

A common source of anxiety for students is that we'll get different answers depending on how we round. In the review section we studied precision and accuracy, but these rules apply to laboratory activities where data comes from actual laboratory measurements that actually have precision and accuracy. Here, like in example 3-17 above, the given information is theoretical and does not come from any actual measurements. Always concerning ourselves with rounding rules is a real hassle, so we use the rule-of-thumb of three significant digits in these academic exercises. But what can create confusion is that we often get somewhat different answers depending on how *often* we round. We'll get slightly different answers if we round every time a mathematical operation is made rather than entering a whole series of calculations into our scientific calculators and rounding only at the final answer. For example in 3-17 above, calculating F_1 and F_2 weights separately, rounding as we go, as opposed to calculating the torque all in one fell sweep in the calculator without rounding each individual step will yield a somewhat different answer. Do not be too concerned about this. When doing theoretical exercises such as in example 3-17, we are not using data acquired from actual measurements and we do not want to overly concern ourselves with precision and accuracy rounding rules. As a result, some variation in our answers is expected. One thing to keep in mind though: don't write answers with more than three significant digits. Excessive, unsubstantiated accuracy is a mathematical error.

RESULTANT TORQUE **L5**

Student Name(s)_____

_____

_____

_____

OBJECTIVE:
Upon completion of this lab, the student will be able to
- define torque, lever arm, and rotational equilibrium.
- describe in words and in equations the conditions necessary for rotational equilibrium.
- calculate torque, rounding appropriately.
- collect data, correctly recording measurements as to their precision and accuracy.
- define the difference between mass and weight.
- experimentally verify the solution to an equilibrium situation.

DISCUSSION:

Torque (τ or sometimes T) is the force-like quantity in rotational systems. Torque causes an object to twist or turn.

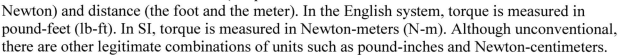

The direction of torque is described as either clockwise (-) or counterclockwise (+).

Torque is equal to the product of the force (F) and the length of the lever arm (l) expressed by the formula
$$\tau = Fl$$
The units of measurement for torque are the product of the units of measurement for force (the pound and the Newton) and distance (the foot and the meter). In the English system, torque is measured in pound-feet (lb-ft). In SI, torque is measured in Newton-meters (N-m). Although unconventional, there are other legitimate combinations of units such as pound-inches and Newton-centimeters.

It is important to understand that the force and lever arm must be perpendicular (intersect at ninety degrees) in order for the formula to work. When the force and lever arm are not perpendicular, trigonometry is used to calculate how much of the applied force is perpendicular. The rest of the force does not cause torque.

Rotational equilibrium happens when torques cancel out. As a result, there is no rotation. In other words, the sum of the different torque values is zero. Here the student will verify this by hanging weights on a meter stick resting upon a fulcrum. Two weights hung on one side cause two aiding torques while one weight hung on the other side causes an opposing torque. These three signed torques should add up to zero when the system is balanced.

APPARATUS:

Meter stick, three knife clips to hang weights from , an extra knife clip to serve at the fulcrum, fulcrum stand, set of slotted weights and hangers.

PROCEDURE:

Part One

1. Weigh the meter stick. Record this value in the data table. (This weight is used in Part Two.)
2. Weigh the hanger clips. These add to the weights hung on the beam.
3. Hang three weights from their clips at different distances from center on a meter stick as shown. The fulcrum support is positioned at beam center.
4. By adjusting the amount of mass and/or distances (lever arms), balance the system (equilibrium).
5. Record the values of the three weights and their corresponding lever arm distances in the data table.
6. Calculate the three torques. Don't forget to add the clip mass. Include a positive or negative sign for each torque to indicate direction. Record the three values in the data table.
7. Calculate net torque by adding the three torques. Record the value.

Part Two

1. Position the fulcrum off beam center, say 10 cm.
2. Adjust the weights and/or lever arms of the hangers to balance the system.
3. Record the values of the three weights and their corresponding distances in the data table.
4. Calculate the additional torque created by the weight of the meter stick by multiplying the distance of the fulcrum. This is from beam center to the weight of the stick. Record this value as τ_4.
5. Calculate net torque by adding the four torques. Record the value.

DATA:

Mass of meter stick _____ kg

Mass of knife clip _____ kg (you may assume they are nearly equal in mass)

Part One

Mass	Weight $F = mg$	Lever Arm	Torque $\tau = Fl$
$m_1 =$ _____ **kg**	$F_1 =$ _____ N	$l_1 =$ _____ cm	$\tau_1 =$ _____ N-cm
$m_2 =$ _____ **kg**	$F_2 =$ _____ N	$l_2 =$ _____ cm	$\tau_2 =$ _____ N-cm
$m_3 =$ _____ **kg**	$F_3 =$ _____ N	$l_3 =$ _____ cm	$\tau_3 =$ _____ N-cm

$$\tau_{net} = \tau_1 + \tau_2 + \tau_3$$

$\tau_{net} = \pm$ _____ $\pm$ _____ $\pm$ _____ $=$ _____ N-cm

Part Two

Mass	Weight $F = mg$	Lever Arm	Torque $\tau = Fl$
$m_1 =$ _____ **kg**	$F_1 =$ _____ N	$l_1 =$ _____ cm	$\tau_1 =$ _____ N-cm
$m_2 =$ _____ **kg**	$F_2 =$ _____ N	$l_2 =$ _____ cm	$\tau_2 =$ _____ N-cm
$m_3 =$ _____ **kg**	$F_3 =$ _____ N	$l_3 =$ _____ cm	$\tau_3 =$ _____ N-cm
$m_4 =$ _____ **kg**	$F_4 =$ _____ N	$l_4 =$ _____ cm	$\tau_4 =$ _____ N-cm

$$\tau_{net} = \tau_1 + \tau_2 + \tau_3 + \tau_4$$

$\tau_{net} = \pm$ _____ $\pm$ _____ $\pm$ _____ $\pm$ _____ $=$ _____ N-cm

FOLLOW UP:

1. Explain in your own words what is meant by *torque*.

2. What is meant by rotational equilibrium?

3. Why was $9.81 m/s^2$ multiplied to the mass?

4. Explain the difference between mass and weight.

5. When the system was balanced, it was obvious no torque rotated the stick. But your calculated value was no doubt not exactly equal to zero. How do you account for this?

6. In this exercise, how many significant digits were the indirectly measured torque values limited to? Why?

7. Explain the sign (±) convention used to define direction of torque.

1. Calculate the torque created by a 20 pound force acting on a 16 inch lever arm.

2. Calculate the torque created by a 120N force acting on a 50 cm lever arm.

3. From the diagram shown, what is the minimum amount of force barely required to move the 200 lb load?

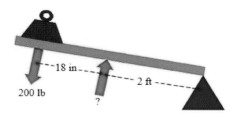

4. From the diagram shown, calculate the torque on the nut.

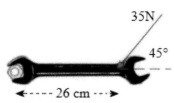

5. From the diagram shown, calculate the torque,

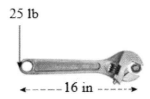

6. From the diagram shown, calculate the resultant torque.

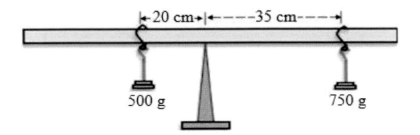

7. For the diagram shown, calculate the resultant torque.

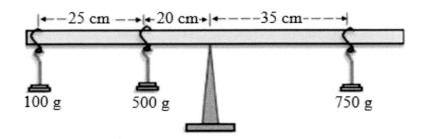

8. Calculate the value of **F** that will balance the system.

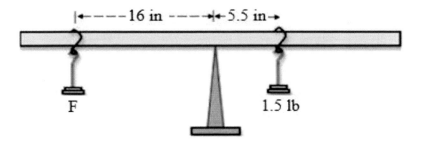

Answers: 1. 320 in-lb or 26.7 ft-lb   **2.** 6000 N-cm or 60.0 N-m   **3.** 350 lb (*This answer is only approximately correct because the 200 lb weight vector always points straight down and will not always be perpendicular to the lever arm as the system rotates.*)   **4.** −6.43 N-m   **5.** 400 in-lb   **6.** −159 N-cm   **7.** −115 n-cm   **8.** 0.516 lb

Following our unifying principle, **pressure** is the mover quantity in both hydraulic and pneumatic fluid systems. Pressure moves fluids much like force moves objects or how voltage moves charge. Pressure is defined as the force per unit area "in" a fluid defined by the formula p = F/A.

Pressure is represented by the letter "p" in formulae, sometimes as a capital letter and sometimes lower case. Here we will keep pressure represented by the lower case "p" in order to distinguish it from the quantity power, represented with capital "P", discussed later. As always, force is represented by the capital "F" and area by capital "A".

> **Fluid Pressure**
>
> $$p = \frac{F}{A}$$

The units of measurement for pressure are units of force divided by units of area. The standard pressure units in the English system is the lb/in^2, or **psi**, but lb/ft^2 is often used in calculations. In the metric system, pressure units are once again units of force divided by units of area. This results in N/m^2 and N/cm^2. The N/m^2 has been named after a famous scientist, and equivalent to a **Pascal**, abbreviated "Pa". The Pascal is a small unit of measure, so the kilopascal, or kPa, is often used.

The use of psi and kPa instead of lb/in^2 and N/m^2 is largely a matter of convenience when writing pressure specifications on labels and plates. For example, "psi" and "kPa" are much more easily labeled and easier to read on the side of a tire than lb/in^2 and N/m^2.

Sometimes technicians will simply say "pounds" when referring to fluid pressure, leaving the "per square inch" unsaid. This commonly used jargon doesn't cause any problems as long as the actual units of pressure are understood.

Pressure in a Cylinder

A **piston** is a device that converts fluid energy into mechanical energy or visa-versa. Since pistons almost always have a circular cross-sectional area, they take the shape of a cylinder. Typically, the term **cylinder** is used instead of "piston" in tech-speak.

The formula p =F/A applies to cylinders. The pressure in the cylinder is the force at the cylinder shaft divided by the area of the cylinder's cross-sectional area, usually circular. Recall the formulae for calculating the area of a circle, where "r" is the radius and "d" is the diameter. Also, the radius is half the diameter, or d = 2r.

> **Area of a Circle**
>
> $$A = \pi r^2$$
> and
> $$A = \frac{\pi d^2}{4}$$

Example 3-18

Calculate the pressure in the cylinder.

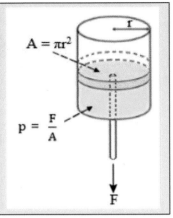

Solution:

$A = \pi r^2 = 3.14(1.5\text{in})^2 = 7.07 \text{ in}^2$

$p = \dfrac{F}{A} = \dfrac{80 \text{ lb}}{7.07\text{in}^2} = 11.3 \text{ lb/in}^2$

Example 3-19

Calculate the force a 2 inch diameter cylinder will produce at 90 psi.

Solution:

$A = \pi r^2 = 3.14(1\text{in})^2 = 3.14 \text{ in}^2$

$F = pA = 90 \text{ lb/in}^2(3.14\text{in}^2) = 283 \text{ lb}$

Example 3-20

Calculate the force produced by a cylinder having a diameter of 10cm at 70kPa.

Solution:

$r = 5\text{cm} = 0.05\text{m}$

$p = 70\text{kPa} = 70{,}000 \text{ Pa} = 70{,}000 \text{ N/m}^2$

$A = \pi r^2 = 3.14(0.05\text{m})^2 = 0.00785 \text{ m}^2$

$F = pA = 70{,}000\text{N/m}^2(0.00785\text{m}^2) = 550 \text{ N}$

Exponential Changes in Cylinders

An interesting thing about cylinders is how a small change in the diameter of the cylinder causes "exponential" (much larger) changes in either the force, the pressure, or both. In $A = \pi r^2$, note there is an exponent of two on "r", so any change in the radius (r) causes exponential changes in the other quantities. In $p = F/A$, the area is in the denominator, so a small *increase in A causes the pressure to decrease exponentially*. Manipulating the formula for F we have $F = pA$, where A is now in the numerator. This means *any small increase in A causes an exponential increase in F*. For example, a circle with a radius equal to one inch has area of 3.13 in². But a circle with a diameter that is three times larger has an area of 7.07 in², nine times more area. Three times more diameter, nine times more area, an exponential (squared) increase.

Blaise Pascal (1623 - 1662)

Blaise Pascal was a French mathematician, physicist, inventor, writer and Christian philosopher. He was a child prodigy who was educated by his father, a tax collector in Rouen. Pascal's earliest work was in the natural and applied sciences where he made important contributions to the study of fluids, and clarified the concepts of pressure and vacuum by generalizing the work of Evangelista Torricelli. Pascal also wrote in defense of the scientific method.

In 1642, while still a teenager, he started some pioneering work on calculating machines. After three years of effort and fifty prototypes, he was one of the first two inventors of the mechanical calculator. He built 20 of these machines (called Pascal's calculators and later Pascalines) in the following ten years. Pascal was an important mathematician, helping create two major new areas of research: he wrote a significant treatise on the subject of projective geometry at the age of 16, and later corresponded with Pierre de Fermat on probability theory, strongly influencing the development of modern economics and social science.

The metric unit of pressure, Newtons per square meter, was named after him.

Density

The density of a substance is the mass of that substance divided by its volume. The greater the density of a material, the more matter is packed into a given space. Density is commonly represented by the Greek letter rho (ρ), written a bit like lower case "p", so make an effort to make these letters look different. Like most lower case Greek letters, rho is written starting at the bottom then swooping up into a loop. It does not have a straight back.

Dividing mass by volume produces the units grams per cubic centimeter (g/cm^3) or kilograms per cubic meter (kg/m^3) in the metric system. In the English system density is measured in slugs per cubic inch ($slugs/in^3$) or slugs per cubic foot ($slugs/ft^3$).

Mass Density
$\rho_m = \dfrac{m}{V}$

Dividing the mass by the volume is the **mass density** of a substance, designated ρ_m.

Example 3-21

Calculate the mass density of a cube 2 cm on a side and having mass of 12.2 grams.

Solution:

$$V = LWD = (2cm)^3 = 8 \text{ cm}^3$$

$$\rho = \frac{m}{V} = \frac{12.2g}{8cm^3} = 1.53 \text{ g/cm}^3$$

Mass Density

When designing the metric system, scientists came up with a very convenient measure of mass. Using water as the standard, they agreed that one cubic centimeter of water should be one gram. As a result, the density of water is defined as being exactly 1 g/cm^3. Also remember that since $1cm^3 = 1cc = 1mL$, then the density of water can also be expressed as 1 g/cc and 1 g/mL.

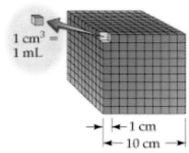

This makes things incredibly straightforward. Imagine a beaker containing 0.5L of water. That half liter is 500 mL, $500cm^3$, 500cc, and since it is water, and water has density of $1g/cm^3$, that beaker contains 500 grams!

One liter of water has one kilogram mass.

Weight Density

Density can also be quantified by using the weight of a substance. The **weight density** of a substance is the weight of the substance divided by its volume. This results in units of pounds per cubic foot (lb/ft^3) in the English system and Newtons per cubic meter (N/m^3) in SI.

Weight Density
$\rho_w = \dfrac{w}{V}$

Density of Water

$$\rho_m = 1 \text{ g/cm}^3 = 1\text{g/cc} = 1\text{g/mL} = 1\text{kg/L} = 1.94\text{slugs/ft}^3$$

$$\rho_w = 62.4 \text{ lb/ft}^3 = 8.34 \text{ lb/gal} = 9.81\text{N/L} = 9810\text{N/m}^3$$

Example 3-22

The tank on a tanker truck contains fifteen thousand gallons of water. How much weight in water is the truck hauling?

Solution:

$$\rho_w = \frac{w}{V}$$

$$w = \rho_w V = 8.34\text{lb/\cancel{gal}}(15,000\cancel{\text{gal}}) = 125,000 \text{ pounds}$$

Specific Gravity

"Specific gravity" is a simple relationship that sounds a bit complicated. It's a strange name for a property that has little to do with gravity. The **specific gravity** of a substance is the density of the substance divided by the density of water. Specific gravity has no conventional literal or abbreviation. We either have to write out "specific gravity" completely or maybe we can agree to use S.G. to save space and pencil lead.

Since specific gravity is a comparison of density to density, the units cancel completely out. These kind of comparison is called a **ratio**. Ratios have no units. Specific gravity is a "pure" number that defines how many times more dense a substance is as compared to water.

Water is the standard. All other substances are compared to water.

Specific Gravity

$$\text{Specific Gravity} = \frac{\text{Density of a Substance}}{\text{Density of Water}}$$

Density and Specific Gravity of Selected Substances		
Solids	**Density**	**Specific Gravity**
Gold	19.3 g/cm^3	19.3
Lead	11.3 g/cm^3	11.3
Silver	10.5 g/cm^3	10.5
Copper	8.9 g/cm^3	8.9
Brass	8.6 g/cm^3	8.6
Steel	7.8 g/cm^3	7.8
Aluminum	2.7 g/cm^3	2.7
Balsa Wood	0.3 g/cm^3	0.3
Oak Wood	0.8 g/cm^3	0.8
Liquids		
Mercury	13.6 g/cm^3	13.6
Seawater	1.04 g/cm^3	1.04
Water	1 g/cm^3	1
Oil	0.9 g/cm^3	0.9
Alcohol	0.8 g/cm^3	0.8

Scanning down the above list of specific gravities you'll notice that in the metric system the numerical value of the specific gravity equals the density of the substance. This is because with water density established at 1 g/cm^3, dividing by one doesn't change things numerically. For example, with mercury's density at 13.6 g/cm^2, this of course means it is 13.6 times denser that water at 1g/cm^3.

In the English system the specific gravity values are the same; S.G. for mercury is still 13.6. Dimensionless (having no units), the specific gravity ratio can be very useful; the number will work with any systems of units.

It might be worth mentioning how incredibly dense gold is, roughly twice as dense as lead!

Example 3-23

The tank on a tanker truck contains fifteen thousand gallons of oil. How much weight in oil is the truck hauling?

Solution:

Specific gravity of oil = 0.9

$$\rho_w = \frac{w}{V}$$

$w = \rho_w V(S.G.) = 8.34\text{lb/gal}(15,000\text{gal})(0.9) = 113,000 \text{ pounds}$

At a specific gravity of 0.9, oil weighs less than water, or .9 the weight of water. As a result, the truck's load is less. Treating the problem as if it were water, then introducing specific gravity is a commonly used method for calculating fluid weights and fluid pressures. Specific gravity is a ratio having no units. For convenience, here we shall use 0.9 for an approximate value for all oils and fuels.

Depth Pressure

Pressure increases with the depth of a fluid because of the fluid's weight density.

In the English system, the weight density of water is 62.4 lb/ft³. In other words, one cubic foot of water weighs 62.4 pounds. This means that at a depth of one foot, 62.4 pounds of weight is pressing (press-ure) down onto one square foot at the bottom, a pressure of 62.4lb/ft². Should we stack another "block" of water, another cubic foot, on top of the first cube, then the weight doubles onto the square foot at the bottom. At a depth of two feet, the pressure is then twice as much or 124.8 lb/ft². The depth pressure is greater for fluids more dense than water, say seawater or mercury, because the weight is greater.

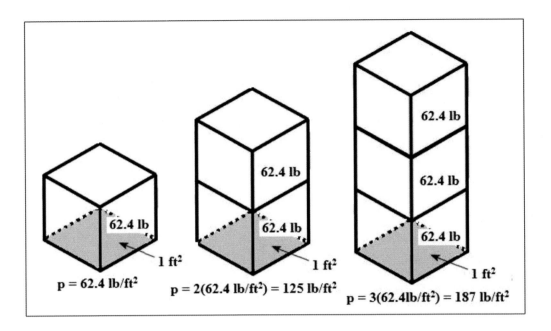

This idea of stacking blocks of fluid atop each other in a column thereby increasing the weight at the bottom of the "stack" is called "a column of fluid". This is a good way to understand how the pressure in a fluid increases with the depth. It is why the following formula is sometimes called the "stack" formula, because depth pressure depends on two things: the weight density of the fluid and the number of blocks that are stacked, the depth.

The depth pressure formula traditionally uses an "h" for height to represent depth. Call it either height or depth, what is important is how tall the "stack" is.

There appear to be two depth pressure formulae at right, but really there is only one. In one of them we multiply the weight density (ρ_w) to the depth (or height, h). In the second the mass density is used, but this must be converted to weight density by multiplying by acceleration due to gravity; it is the weight of the fluid that causes depth pressure.

Depth Pressure
$p = \rho_w h$ or $p = \rho_m g h$

Example 3-24

Calculate the pressure (in psi) at the bottom of a swimming pool ten feet deep.

Solution:

$$p = \rho_w h = 62.4 \text{lb/ft}^3 (10 \text{ft}) = 624 \text{ lb/ft}^2 = 4.33 \text{ psi}$$

Careful: $1ft^2 = 144 in^2$

One handy relationship often used by technicians, especially plumbers, is that pressure increases by 0.433 psi every foot, or 0.433 psi/ft. Dividing by twelve we also then have 0.0361 psi/in, which is useful in fluid-related instruments such as manometers discussed below.

It should be mentioned that here 0.433 psi/ft has been rounded down to three digits but, as in most applications, plumbing technicians round it up to 0.434 psi/ft, preferring to be over in their pressure calculations rather than slightly under. This is the value they use for calculations on their licensing exams.

Water pressure increases with depth at a rate of 0.433 psi/ft or 0.0361 psi/in.

Example 3-25

Calculate the pressure (in psi) at the bottom of a swimming pool ten feet deep.

Solution:
$$p = 0.433 \text{ psi/ft}(10 \text{ ft}) = 4.33 \text{ psi}$$

Example 3-26

Calculate the pressure (in psi) in the ocean at a depth of 2800 feet.

Solution:
$$p = \rho_w h(S.G.) = 62.4 \text{lb/ft}^3(2800\text{ft})(1.04) = 182{,}000 \text{ lb/ft}^2 = 1260 \text{ psi}$$

Alternate solution:
$$p = 0.433 \text{psi/ft}(2800\text{ft})(1.04) = 1260 \text{ psi}$$

Specific gravity is multiplied into the equation because seawater is on average 4% heavier than pure water, so the pressure is greater.

Example 3-27

Calculate the pressure (in kPa) at the bottom of a swimming pool six meters deep.

Solution:
$$p = \rho_w h = 9810 \text{N/m}^3(6\text{m}) = 58900 \text{ N/m}^2 = 58900 \text{ Pa} = 58.9 \text{ kPa}$$

Atmospheric, Gauge, and Absolute Pressure

Gasses are fluids that have weight just like liquid. And like any liquid, the air around you, having weight, causes pressure that increases with depth. **Atmospheric pressure** is the pressure created by the weight of the air around us. If you're at sea level in ordinary weather conditions, the pressure on your body is 14.7 psi (101kPa). The **atmosphere (atm)** is a unit of measurement equal to standard atmospheric pressure at sea level.

> **Atmospheric Pressure**
>
> p_{atm} = 1 atm =14.7 psi = 101 kPa

To a large extent, gasses and liquids are different in only one significant way: while liquids are virtually incompressible (meaning the density does not change easily) gasses are easily compressed, so the density of a gas can change dramatically as pressures change. This is why the density of the air is greater at lower elevations and less at higher elevations. For example, one of the greatest challenges of those climbing Mt. Everest is dealing with the "thin" air at those high altitudes. It is also why passenger jets must have pressurized cabins. As a result, the stack formula cannot work for atmospheric pressure. Unlike liquids, say water, where the weight increases linearly 62.4 pounds for every foot of depth, the air density does not change at a constant rate with depth.

Gage pressure is the pressure measured by a gage. This might seem like a bit of a joke, but there's an important distinction to understand here. Take for example a completely flat tire. How much pressure is in the tire? Zero psi? If we were to apply a pressure gage to the stem, the gage would certainly read 0 psi. This is the gage pressure. But isn't there 14.7 psi in there? Yes. There is also 14.7 psi outside of the tire cancelling it out and preventing the tire from inflating, but the pressure is there. Now let's pressurize the tire to 30 psi. The gage pressure is 30 psi and the atmospheric pressure is 14.7 psi. So starting with 14.7 psi in the tire we added another 30 psi. This means the total pressure in the tire is 44.7 psi, the sum of the gage pressure and the atmospheric pressure. It seems that atmospheric pressure is always around to begin with. Total, or **absolute pressure**, is the sum of the gage pressure and the atmospheric pressure.

> **absolute pressure = gage pressure + atmospheric pressure**
>
> $p_T = p_g + p_{atm}$

> **Example 3-28**
>
> Find the total pressure in a tire with a gage pressure of 32psi.
>
> Solution:
>
> $p_T = p_g + p_{atm}$ = 32psi + 14.7psi = 46.7psi

Manometers

A **manometer** is a gage used to measure differential pressure, that is, the difference in pressure between two points. It is a simple, inexpensive, and very reliable device that has no mechanical parts. Because of its reliability, it is often used in medical procedures. It is used to calibrate other mechanical pressure gages.

A manometer is a U-shaped tube containing a fluid. The fluid can be ordinary water, but the fluid is usually a colored liquid specifically designed for use in manometers, a liquid called "gage fluid". When there is a pressure difference between the two openings, or "ports", at the top, the fluid in the tube is displaced to a point a little higher level on one side. This difference in height is directly proportional to the pressure difference across the ports. Using the "stack" formula, the pressure can be reliably determined.

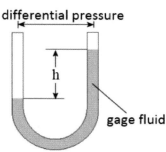

When measuring high pressures, water manometers can get unmanageably tall, so mercury, a much heavier fluid, is used because with its greater weight it does not move as far up the manometer column.

We have discussed numerous units of measurement used to quantify pressure, but here's a new one. Remember that pressure increases consistently with depth, that for water, each foot corresponds to 0.433 psi and each inch corresponds to 0.0361psi. When using manometers, the pressure is measured in inches or centimeters height on the tube. But since we know each inch represents a particular pressure for a particular fluid, why not just say "inches of water" or "inches of mercury" and leave it at that? This is how pressure is often measured, by simply stating the height in inches or centimeters of a particular fluid. This is what the weather man is saying when he says the atmospheric pressure is "30 inches of mercury and rising"; he's saying there's literally thirty inches of mercury of differential height in his gage and that a high pressure area is on the way in to the area.

Mechanical low pressure gages such as the **magnehelic** pressure gages display pressure values in inches of water.

Care should be taken to understand the difference between a manometer and a **barometer**. While manometers can measure any pressure difference across their two ports, a barometer is a pressure gage designed specifically to measure atmospheric pressure.

Example 3-29

A mercury manometer has a 15 inch height difference of fluid in the manometer arms. Calculate the differential pressure in psi.

Solution:
p = 0.0361psi/~~in~~(15~~in~~)(13.6) = 7.36 psi

0.0361psi/in is the height-pressure relationship for water, so this was multiplied by the specific gravity of mercury. At 13.6 times heavier, it takes 13.6 times more pressure to move mercury the same height as water.

Example 3-30

A water manometer has a differential height of 12.5 inches.

What is the gage pressure (in inches), atmospheric pressure (in inches), and absolute pressure (in psi).

Solution:

$P_g = 12.5$ in

$P_{atm} = 14.7psi = 14.7psi\left(\dfrac{1in}{0.0361psi}\right) = 407$ in

$P_T = P_g + P_{atm} = 12.5in + 407in = 419.5$ in

$P_T = 419.5\ in\left(\dfrac{0.0361psi}{1in}\right) = 15.1$ psi

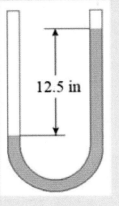

12.5 in

Buoyancy

Any object immersed in a fluid has an upward force acting on it called the **buoyant force**. Some objects float, but even the ones that sink are buoyed up a bit, lighter.

In an earlier chapter we discussed fluid volume. To understand buoyancy it is first necessary to understand what is meant by a "volume of fluid displaced". If an object is placed in a container filled right to the brim with fluid, some of the fluid will overflow when an object is dropped into the fluid. This amount of fluid that overflowed is the **volume of fluid displaced** by the object. If the container was not filled with fluid to the brim, the level of the fluid in the container would rise. This change in the height is also a measure of the displaced volume of fluid. And here's the important connection: the volume of fluid displaced equals the volume of the object displacing the fluid.

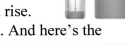

*If you dip your toe in the water, it's **immersed**. If you jump in and sink, and immersion is total, you're **submerged**.*

A completely submerged object always displaces a volume of fluid equal to its own volume.

The relationship between buoyancy and displaced fluid was first discovered by Greek scientist Archimedes the third century BC:

An immersed object is buoyed up by a force equal to the weight of the fluid it displaces.

This relationship between buoyancy force and displaced fluid is **Archimedes' principle**. The buoyancy force is equal to the weight of the fluid displaced.

Example 3-31

Two solid blocks of identical size are submerged in water. One block is made of lead and the other block is made of aluminum. Upon which is the buoyancy force greater?

Answer:
The buoyancy force is the same on each. Equal in size, they each displaced the same volume of water.

Example 3-32

A cube 6 cm on a side is submerged in water. Calculate the buoyancy force.

Solution:
$V = s^3 = (6cm)^3 = 216 \text{ cm}^3$

$\rho_m = \dfrac{m}{V}; \quad m = \rho_m V = 1g/cm^3(216cm^3) = 216 \text{ g}$ *(Could have been done by inspection.)*

$w = mg = 0.216kg(9.81m/s^2) = 2.12N$

Buoyancy is a force measured in Newtons.

Example 3-33

An 11 pound cube 6 inches on a side is submerged in seawater. Calculate the buoyancy force and the weight of the object when submerged.

Solution:
$V = s^3 = (0.5ft)^3 = 0.125 \text{ ft}^3$

Buoyancy: $\rho_w = \dfrac{w}{V}; \quad w = \rho_w V = 62.4 \text{ lb}/ft^3(0.125ft^3)(1.04) = 8.11 \text{ lb}$

Weight of submerged object = 11 lb – 8.11 lb = 2.89 lb

The weight density of water must be multiplied by the specific gravity of seawater because seawater, being heavier, produces more buoyancy.

Hydrometers

Some objects don't sink completely in a fluid, they float. This means that the object is less dense than the fluid. In other words, the fluid is heavy enough to create a buoyancy force at least equal to the weight of the object. This is why wood floats in water, lead floats in mercury, and helium balloons rise in air.

The floating object sinks down only to the point of displacing enough fluid equal to its weight. This is the principle upon which hydrometers work. A **hydrometer** is an instrument that measures the specific gravity of a liquid. Hydrometers are used widely in industry from brewing beer to refining fuels. They are used for four primary purposes: First, they're used to monitor the progress of certain liquid chemical reactions. Second, they're used to determine the concentration of liquid mixtures. Third, they're used to automatically control the blending of lube oils. And fourth, they're used to measure the change in density of a liquid as it changes with changing temperatures.

The hydrometer is a glass bulb that has a weight at one end and a scale at the other. The scale is marked off (calibrated) to show a range of specific gravity values. In the metric system these specific gravity values are numerically equal to the mass density of the liquid in g/cm^3.

When measuring the specific gravity of a liquid, the hydrometer floats, weight down, in the liquid, and floats at a particular height depending on the weight in the bulb and the density of the fluid. According to Archimedes' principle, an upward buoyancy force equal to the weight of the displaced fluid will float the hydrometer (as long as the hydrometer isn't too heavy). The specific gravity of the liquid is the reading on the scale that's level with the surface of the liquid.

> **The hydrometer sinks into the fluid until it displaces a volume of water equal to its weight, so it then floats at a particular height in the fluid.**

Archimedes (287 BC – 212 BC)

Archimedes of Syracuse was an Ancient Greek mathematician, physicist, engineer, inventor, and astronomer. Although few details of his life are known, he is regarded as one of the leading scientists in classical antiquity.

Generally considered the greatest mathematician of antiquity and one of the greatest of all time, Archimedes anticipated modern calculus and analysis by applying concepts of infinitesimals and the method of exhaustion to derive and rigorously prove a range of geometrical theorems, including the area of a circle, the surface area and volume of a sphere, and the area under a parabola. Other mathematical achievements include deriving an accurate approximation of pi, defining and investigating the spiral bearing his name, and creating a system using exponentiation for expressing very large numbers. He was also one of the first to apply mathematics to physical phenomena, founding hydrostatics and statics, including an explanation of the principle of the lever. He is credited with designing innovative machines, such as his screw pump, compound pulleys, and defensive war machines to protect his native Syracuse from invasion. Archimedes'' Principle of buoyancy is of course named after him.

Archimedes died during the Siege of Syracuse when he was killed by a Roman soldier despite orders that he should not be harmed. Cicero describes visiting the tomb of Archimedes, which was surmounted by a sphere and a cylinder, which Archimedes had requested to be placed on his tomb, representing his mathematical discoveries.

1. Explain in your own words why fluid pressure is considered a force-like mover quantity.

2. Why does fluid pressure increase with the depth in a fluid?

3. Find the pressure at the bottom of a swimming pool that is 12 feet deep.

4. A beaker contains 250 mL of water.
    a. How many ccs of water is there in the beaker?

    b. How many grams of water is there in the beaker?

    c. What is the weight of the water in the beaker?

5. Find the pressure at the bottom of a lake that is 12.0 meters deep.

6. A vertical pipe (stack) has a leak somewhere along its vertical height. The pressure at the bottom of the stack is 6.20 psi. At what height is the leak?

7. Identify the material of a cube by calculating the density. The cube's mass is 15.4 grams and it is 1.20 cm on each side.

8. A 50.0 pound force is applied to the shaft of a 2.00 inch diameter cylinder. Calculate the pressure produced in the cylinder.

9. What does a manometer measure?
   How does it work?

10. What is the total pressure in a tire when the tire pressure gage reads 30.0 psi?

11. A u-tube mercury manometer shows a height differential of 10.5 inches. What is the pressure in psi?

12. What is the depth pressure 1.00 mile down into the ocean?

13. A hydraulic lift is used to lift vehicles in an automotive repair shop. Shown at left, hydraulic pressure is produced on the smaller piston at left. This pressure is transmitted to the larger piston at right to increase force. If 20.0 lb is applied to the 4.00 inch diameter smaller piston, how much force can be produced at the 36.0 inch diameter larger piston?

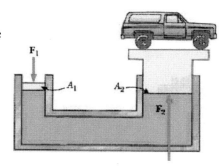

14. Explain in words Archimedes' Principle as it applies to buoyancy.

15. An object weighs 20.0 pounds. When this object is immersed in water, five pounds of water is displaced.

    a. What is the buoyancy force?

    b. How much does the object weigh while in the water?

    c. Should the object have been immersed in seawater instead of pure water, would the buoyancy force be different? Why?

16. A u-tube water manometer shows a difference in height of 18.2 inches. What is the differential pressure in psi?

17. What is the buoyancy of a 10.0 cubic centimeter chunk of aluminum?

18. A piston has a diameter of 3.00 inches. How much force can it produce at 160 psi?

19. At what pressure will a 5.00 inch diameter piston produce 2500 lb?

20. A simple u-tube manometer is filled with mercury (density of mercury equals 0.490 lb per cubic inch) is connected between two bottles of gas with different pressures. The difference in height between the two columns in the manometer is 8.40 inches. What is the difference in pressure between the two manometer ports?

Answers:
1. pressure moves fluids much like force moves objects   **2.** as the depth increases so does the amount of fluid and therefore also the weight pressing down **3.** 749 lb/ft² or 5.20 psi **4a.** 250 cc **4b.** 250 g **4c.** 2.45N **5.** 118 kPa **6.** 14.3 ft **7.** copper **8.** 15.9 psi **9.** measures differential pressure, the difference in height in the arms is proportional to the difference in pressure across the two openings, depth pressure **10.** 44.7 psi **11.** 5.16 psi **12.** 2380 psi **13.** 1620 lb **15a.** 5 lb **15.b** 15 lb **16.** 0.660 psi **17.** 0.0981N **18.** 1130 lb **19.** 128 psi **20.** 4.13 psi

Student Name(s) _____

_____

_____

Objective:

Upon completion of this lab, the student will be able to

- define pressure as it relates to pistons.
- calculate the cross-sectional area of a cylinder given its diameter.
- calculate the pressure in the piston given the force applied to its shaft along with cylinder dimensions.
- explain in words the difference between mass and weight.
- explain in words the relationship between force, pressure, and head area in a cylinder and how any of the three quantities changes with respect to the others

Discussion:

A piston is a hollow cylinder within which a piston head moves a piston shaft when fluid in the cylinder is under pressure. In hydraulic cylinders, a liquid such as oil is the fluid under pressure. In pneumatic cylinders, the operating fluid is a gas such as air. Here a pneumatic cylinder is used.

In this lab students will hang a weight on the shaft. This will compress the air under the piston head and create pneumatic pressure. The student will then calculate the pressure given the weight (force) and piston size, comparing this calculation to a pressure gage.

Three quantities are to be explored here. The pressure (p), the piston head area (A), and the force on the shaft (F) interact in this system. Their relationship is expressed by the formula

$$p = \frac{F}{A}$$

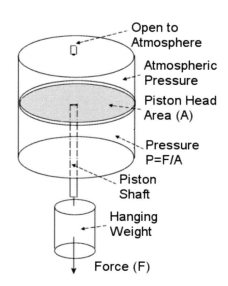

The area of the piston head is usually circular, so given the diameter of the cylinder the area of the head is calculated using

$$A = \pi r^2$$

where π represents exactly the ration of the circumference to the diameter of all circles, approximated 3.14, and r is the internal radius of the cylinder. The diameter of the cylinder is twice the radius, so another way of calculating the area given the diameter is

$$A = \pi d^2 / 4$$

In the English system of units, the force is usually measured in pounds (lb) and the area in square inches (in^2). When these units are divided, the resulting units of pressure become pounds per square inch or psi.

In the metric system of units, the force is usually measured in newtons (N) and the area in square meters (m^2). When these units are divided the resulting units of pressure are N/m^2. 1 N/m^2 = 1 Pascal (Pa)

Generally the load on the shaft is recorded as kilograms (kg). This is a measure of mass, not force. Mass is a measure of how much matter there is in an object. Weight is the force created when gravity acts on the mass. The pressure formula requires units of force in either pounds or newtons. Remember that in order to have units of force, kilograms must be multiplied by 9.81 to convert to Newtons in the metric system, and when using the English system, kilograms should be multiplied by 2.20 lb/kg to convert to pounds.

The formula $p = \dfrac{F}{A}$ can be manipulated into $F = pA$ and $A = \dfrac{F}{p}$.

A very useful and simple way of understanding relationships in any formula is by noting which quantities are either directly or indirectly proportional to the isolated letter. For example, in $p = F/A$ the pressure is directly proportional to F because F is in the numerator on the other side of the equation (an increase in F causes an increase in p). Conversely, A is in the denominator on the other of the equation, so A is indirectly proportional to P (an increase in A causes a decrease in p). This direct verses indirect method understanding relationships between quantities in a formula can be applied to any formula.

APPARATUS:

Piston assembly ($1\frac{1}{16}$ inch diameter)
Heavy duty weight set
Pressure gage (0-30psi)
S hook
C-clamps

Procedure and Data:

1. Clamp the apparatus securely to the edge of a table or work bench as shown.

2. Unless otherwise specified, the piston diameter should be one $1\frac{1}{16}$ inch.

   $D = 1\frac{1}{16}$ inch.

3. Determine the radius of the piston by using the simple formula $r = D/2$.

   r = _____ in

4. Calculate the cross-sectional area of the piston by using the formula $A = \pi r^2$.

   A = _____ in^2

5. With the pressure gage removed (allows air to move freely) move the piston shaft at its uppermost position.

6. Connect the pressure gage.

7. Hang enough mass from the piston shaft so that an approximately center-scale reading appears on the pressure gage.

   m = _____ kg

8. Convert this mass in kilograms to pounds, using 1 kg = 2.2 lb.

   F = _____ lb

9. Using the formula $P = F/A$, calculate the pressure in the cylinder in psi.

   p = _____ psi (calculated)

10. Read the pressure gage and record the value.

    p = _____ psi (gage)

Follow Up:

1. What happens to the pressure in a piston if the force is increased?

2. What happens to the pressure in a piston when the area is increased?

3. What happens to the force in a piston when the area is increased?
   (Manipulate the formula $p = F/A$ for F)

4. What is the difference between mass and weight?

5. Why was the mass multiplied by 2.2?

6. By how much would the cross-sectional area change if the diameter were tripled? (Careful! The area increases exponentially with the diameter.)

7. How much weight would the piston in this lab be able to lift when 20psi is applied? Please write a complete solution as shown in class.

Student Name(s) _____

_____

_____

Measuring Differential Pressure with a Manometer

Objectives:

Upon finishing this lab, the student will be able to…
- explain in words what a manometer is and how it works.
- explain in words what atmospheric pressure is and what causes it.
- explain the difference between atmospheric pressure, absolute pressure, and gage pressure.
- use a manometer to measure pressures above and below atmospheric pressure.
- compare pressures measured with a manometer to that measured with a magnehelic gage.
- explain in words how pressure can be measured in units of inches or centimeters of a fluid.

Main Ideas:

- Manometers measure differential gage pressure between two points in a system.
- Absolute (or total) pressure is the gage pressure added to atmospheric pressure.
- Atmospheric pressure is caused by the weight of the air surrounding us.
  $P_{atm} = 14.7 psi.$
- Pressure increases with the depth of a fluid because of the weight of the fluid. Since differential pressure is often measured with instruments utilizing a displaced fluid, fluid pressure is often expressed as a depth (or height) of that fluid.
- Since the weight density of water is $62.4 \, ^{lb}/_{ft^3}$, water pressure increases $62.4 \, ^{lb}/_{ft^2}$ $\left(or \, 0.433 \, ^{lb}/_{in^2} \right)$ for every foot of depth (or height).
- Because of the weight density of water, every inch of water represents 0.0361 psi.

Manometers are used widely in industry to measure gage pressures and pressure differences between two points in a system. Simply a u-tube with fluid in it, manometers have no moving parts, so provide a simple and extremely reliable way to measure differential pressures. For example, mechanical pressure gages are often calibrated using a manometer. Other applications include monitoring pressures across filters and flow rates through pipe lines.

Total pressure equals Gage Pressure plus atmospheric pressure. Atmospheric pressure is caused by the weight of the air around us. Any changes in this pressure, whether added (positive) or taken away (negative) is gage pressure. The sum of these is the total pressure present in a system. Manometers measure gage pressure. For example, before any pressure is added to a tire there is 14.7 psi of atmospheric pressure present in the tire. If the tire is inflated to 30 psi, the gage pressure is 30 psi but the absolute pressure is 30psi + 14.7psi = 44.7 psi. This relationship is represented by the formula: $p_T = p_g + p_{atm}$

A manometer is a u-tube with fluid in it. A difference in pressure across its two ports will cause the fluid to be at different heights in the two arms. When the weight density of the fluid is known, the pressure can then be calculated.

The amount of pressure in a system is often measure in units if psi or kPa. But pressures are also often measured in inches or centimeters of a fluid, with the two primary fluids being mercury (Hg) and water (H_2O). This way of measuring pressure is comes from the fact that fluid pressure increases with the depth of a fluid because of the fluid's weight. The manometer operates on this principle. A dense and heavy fluid, mercury is used to measure large pressures because it takes more pressure to lift it. For smaller pressures, water is often used because it moves more. Since the specific gravity of mercury is 13.6 (it is 13.6 times more dense than water), a mercury manometer can measure pressures 13.6 times larger without a change in height in the manometer arms. As a result, a physically smaller manometer can be used when mercury is the operating fluid.

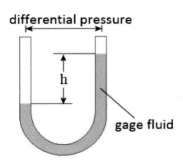

Equipment:

U-tube manometer

60 cc syringe serving as pressure source

Dwyer Magnehelic differential pressure gage

Vinyl tubing and T-connector for connections

12 inch ruler or yard stick

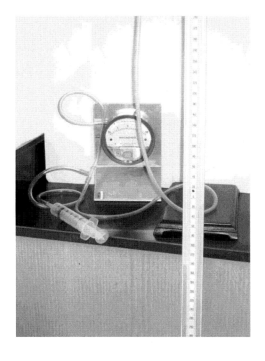

Procedures:

Caution! Be careful no gage fluid is injected into the magnehelic!

Part 1: Positive Pressure

1. Connect the "T" and tubing between the manometer and the magnehelic as shown such that the same pressure is applied to both instruments for comparison.

2. Move the syringe plunger to the 30 cc center position such that positive or negative pressure can be applied, then connect the syringe to the hose.

3. Push the syringe plunger in, creating positive pressure, until the needle on the magnehelic pressure gage reads 5 inches H_2O.

4. Compare the magnehelic gage reading to the manometer by measuring the difference in height between the two water columns in the manometer with a ruler. A positive number represents positive pressure, or pressure above atmospheric. Record the value in the data table.

5. Push the syringe plunger in, creating more positive pressure, until the needle on the pressure gage reads 10 inches H_2O. Recording the value in the data table.

6. Measure the difference in height between the two water columns in the manometer with a ruler. A positive number indicates positive pressure, or pressure above atmospheric. Record the value in the data table.

Part 2: Negative pressure

7. Pull the syringe plunger out to create negative pressure. The fluid in the manometer tube should move the other way. Pull the syringe until the needle on the magnehelic pressure gage reads -5 inches H_2O.

8. Measure the difference in height between the two water columns in the manometer with a ruler. A negative number represents negative pressure, or pressure below atmospheric. Record the value in the data table.

9. Pull the syringe plunger out to create more negative pressure until the needle on the pressure gage reads -10 inches H_2O.

10. Measure the difference in height between the two water columns in the manometer with a ruler. A negative number represents negative pressure, or pressure below atmospheric. Record the value in the data table.

Calculations:

1. Convert atmospheric pressure (14.7 psi) to inches H_2O using the conversion 1 inch H_2O = 0.0361 psi. Record this value in the table for the entire column entitled "Atmospheric Pressure (inches H_2O)".

2. For each run in the table, calculate the absolute (or total) pressure in inches H_2O by adding gage pressure to atmospheric pressure. Record these values in the table.

3. Convert the absolute pressure from units of inches H_2O to psi by using the conversion factor 1 inch H_2O = 0.0361 psi. Record these values in the table.

Data Table:

$$p_g \quad + \quad p_{atm} \quad = \quad p_T$$

Run	Magnehelic Gage Pressure (inches of water)	Manometer Gage Pressure (inches of water)	Atmospheric Pressure (inches of water)	Total Pressure (inches of water)	Total Pressure (psi)
1	+ 5				
2	+10				
3	−5				
4	−10				

Questions:

1. How well did the magnehelic readings match the ruler readings on the manometer?

2. If the manometer and magnehelic pressure readings don't match, which of the two readings is the more correct? Explain why.

3. Explain why pressure increases with the depth of a fluid.

4. In the lab, positive pressure was applied in runs one and two, and negative pressure in runs three and four. Explain why the absolute pressure in psi would be just above 14.7 psi in runs one and two and just below 14.7 psi in runs three and four.

5. Why would mercury be used to measure very large pressures instead of water?

STUDENT CHALLENGE:

Consider the specific gravities of water and mercury. Taking this into account, how tall would the arms of a water manometer have to be to measure a pressure difference of 35 inches Hg?

HYDROMETERS **L8**

Student Name(s)_____

_____

_____

Measuring Specific Gravity with Hydrometers

Objective:

Upon completion of this lab, the student will be able to

- define specific gravity.
- describe what a hydrometer and how it works.
- define buoyancy.
- measure the specific gravity of a liquid with a hydrometer.

Main Ideas:

- The specific gravity of a substance is a number equal to the ratio of the density of the substance divided by the density of water.
- A ratio, specific gravity has no units. The units of density cancel.

Through the principle of buoyancy, hydrometers measure specific gravity of fluids by floating at a certain height in a fluid. A hydrometer is an instrument that measures the specific gravity of a liquid. Hydrometers are used widely in industry from brewing beer to refining fuels. They are used for four primary purposes: First, they're used to monitor the progress of certain liquid chemical reactions. Second, they're used to determine the concentration of liquid mixtures. Third, they're used to automatically control the blending of lube oils. And fourth, they're used to measure the change in density of a liquid as it changes with changing temperatures.

The image at right shows a hydrometer. The hydrometer is a glass bulb that has a weight at one end and a scale at the other. The scale is marked off (calibrated) to show a range of specific gravity values in g/cm^3.

When measuring the specific gravity of a liquid, the hydrometer floats, weight down, in the liquid, and floats a particular height in the liquid depending on the weight in the bulb and the density of the fluid. **Buoyancy** is an upward force on any immersed object. The amount of buoyancy force is equal to the weight of the fluid displace. In the hydrometers, this volume of water is equal to how much of the hydrometer sinks, or the level at which it floats. ***The*** *hydrometer sinks into the fluid until it displaces a volume of water equal to its weight, so it then floats at a particular height in the fluid.*

The specific gravity of the liquid is the reading on the scale that's level with the surface of the liquid. Figure 1 shows a reading of 1.4. Having no units, this measurement of specific gravity means that this liquid is 1.4 times denser than water, a specific gravity of 1.4.

There are different types of hydrometers that are used for certain types of liquids. A commonly used storage battery tester has a small encased hydrometer to determine the acid-water mixture in a battery.

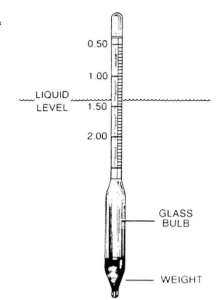

Shown at right, another commonly used type of pocket hydrometer is used to measure antifreeze concentrations in the cooling system of automobiles. This type typically uses four or five colored balls of various densities. The denser (or more concentrated) the mixture, the more balls will float. A rough measurement of the concentration can be determined by counting the number of floating balls. A scale on the side of the hydrometer typically expresses the corresponding freezing point of the mixture to the number of floaters.

APPARATUS:

Set of scaled heavy liquids hydrometers

Set of scaled light liquids hydrometers

Pocket-type hydrometer with colored floatation balls

Four graduated cylinders, at least 10 inches tall

Liquids to mix: tap water, isopropyl alcohol, RV antifreeze

Tongs

PROCEDURE:

1. In the tall graduated containers, prepare the four test liquids: isopropyl alcohol, tap water, and the mixtures of tap water and antifreeze.
2. Carefully place an appropriately scaled hydrometer in one of the test liquids such making sure the hydrometer floats at a height such that a reading from its scale can be determined. If the wrong hydrometer is used, it will either float too high or sink to the bottom. In this case, simply remove the hydrometer (with the tongs if it sinks), rinse and wipe it off and select another hydrometer.
3. In g/cm^3, read the value from the hydrometer scale and record it in the data table provided. Always rinse and wipe the hydrometers when they are removed from the liquid.
4. Use the pocket-type hydrometer with the colored balls to estimate specific gravity by placing the sampling tube into each liquid, squeezing the bulb and releasing to draw fluid. Note in the data table which, if any, colored balls float. Rinse out the pocket hydrometer thoroughly after each use.

DATA:

Test Liquid	Scaled Hydrometer	Pocket Hydrometer
Isopropyl Alcohol		
Tap Water		
50% RV Antifreeze		
100% RV Antifreeze		

Hydrometers Wrap-Up:

1. What is meant by the specific gravity of a substance?

2. Why did some of the colored balls in the pocket-type hydrometer float and others sink?

3. List the probable order of the colored balls in the pocket-type hydrometer according to their densities (Greatest to least).

4. When you look at the scale of a hydrometer, do you find larger values or smaller values of specific gravity at the *lower end*? Explain.

5. Explain why some of the hydrometers float higher in a fluid than others **in terms of Archimedes' Principle**.

Our unifying principle states that each energy system has a mover quantity that causes some sort of displacement. In electrical systems the mover quantity is **voltage**. Voltage moves charge in much the same way that fluid pressure moves fluids. Voltage is symbolized in formulae with either capital "V" or capital "E" (for electromotive force). Here we will use "E" to avoid confusion with volume.

Electric circuit diagrams that use special symbols for circuit components are called **schematic** diagrams. The most common source for DC voltage is the battery. The schematic symbol for a battery is shown at right. A battery has a positive and a negative side. Through a chemical process, electrons bunch up at one end of the battery, the negative side, while at the positive side electrons are missing. The positive side is the end with the longer line. The negative sign, with its shorter line, sort of looks like a negative symbol. The direction of the electrical current from the battery has two conventions. The first is electron flow, where electrons are imagined exiting the negative end of the battery where they have accumulated, and rush to the positive side through a conductor. **Electron flow** is from the negative to the positive terminals. The second convention has been historically the way current flow direction is defined, from positive to negative. Here, we will use the traditional **conventional current flow** direction where current is said to flow from the positive side of the battery to the negative, with negative considered "ground".

Batteries add and subtract like forces. When we stack batteries in the handle of a flashlight, they are inserted such that they are **series aiding**, that is, all push current in the same direction together. In series aiding, battery voltages add up directly. But batteries can also be connected **series opposing**, when the voltages cancel.

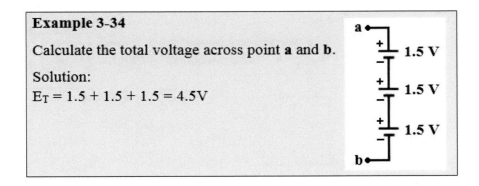

Example 3-34

Calculate the total voltage across point **a** and **b**.

Solution:

$E_T = 1.5 + 1.5 + 1.5 = 4.5V$

Example 3-35

Calculate the voltage across points **a** and **b**.

Solution:

$E_T = 1.5 - 1.5 - 1.5 - 1.5 = 3.0V$, positive at point a

The top battery is opposing the remaining three.

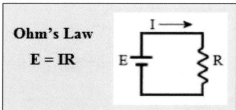

1.5 V
1.5 V
1.5 V
1.5 V

Resistance

Electrical **resistance** holds back, or opposes, electron flow in electric circuits.

In review, the mover in electrical systems is the voltage, the displacement is the charge (q) measured in coulombs, and the rate is the current (I) measured in amperes. Resistance limits the current. Capital "R" is used to represent resistance in formulae.

The relationship between the voltage, current, and resistance is known as **Ohm's Law**, named after the man who first developed it. It is he who the units of resistance is named after, the Ohm. The symbol used to represent the unit Ohm is the Greek upper case omega (Ω). Why not simply "O" for Ohms? Because it can be confused with a zero and "omega" sounds a bit like the man's name. Remember lower case omega (ω) is already used to represent angular velocity.

A **resistor** is an electronic component. It is not a complicated device; it simply holds back the current, dissipating energy by heating up a bit. Resistors wastefully use up energy by releasing the energy as heat. This type of energy release is called **energy dissipation**.

In electronic schematic diagrams resistors are symbolized by a zig-zag, as if the current is having trouble getting through.

Ohm's Law states that the current in a circuit is directly proportional to the voltage applied but inversely proportional to the resistance limiting it. Ohm's Law is commonly expressed as a product: the voltage is equal to the current times the resistance, or $E = IR$.

Ohm's Law

$E = IR$

Example 3-36

Apply Ohm's Law to calculate the current.

Solution:

$$I = \frac{E}{R} = \frac{12V}{600\Omega} = 0.02A = 20mA$$

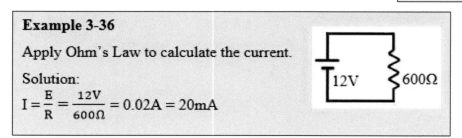

12V   600Ω

Series verses Parallel Circuits

The **series circuit** shown is called a "single loop" or "series" circuit because there is only one path for current to take as it flows from one terminal of the battery to the other. By contrast, in the **parallel circuit**, there are three basic loops – the current splits, with the three resistors sharing the total current coming from the voltage source. A parallel circuit is when there are multiple paths for the current.

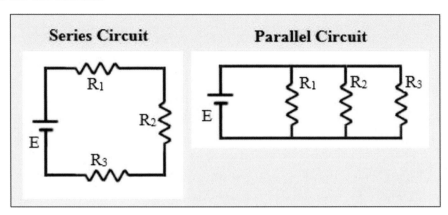

In the series circuit the current is the same in each resistor, but the resistors share the voltage. Parallel circuits are the other way around. In the parallel circuit, the voltage is the same across each resistor but the total current splits and is shared.

Series Circuits	Parallel Circuits
$I_T = I_1 = I_2 = I_3$	$I_T = I_1 + I_2 + I_3$
$V_T = V_1 + V_2 + V_3$	$V_T = V_1 = V_2 = V_3$

Calculating Total Resistance

The voltage is the mover in electrical systems. The voltage source "pushes" electrons through the resistors. The total resistance the voltage source "feels" not only depends on the individual values of each resistor, but also how they are connected. The voltage source "feels" only the total resistance (R_T) of the circuit. Total resistance is calculated depending on how the resistors are connected.

In series circuits, the total resistance is simply the *sum total* of all the value of the resistors. In parallel circuits the resistance is considerably less because the current has more paths it can take. There is always less resistance when resistors are connected in parallel. With current increasing with each parallel branch, it is the reciprocals of the resistances that add. The total resistance of a parallel circuit can be calculated using the so-called "reciprocal formula".

Series Resistors	Parallel Resistors
$R_T = R_1 + R_2 + R_3$	$\dfrac{1}{R_T} = \dfrac{1}{R_1} + \dfrac{1}{R_2} + \dfrac{1}{R_3}$
	$R_T = \dfrac{R_1 R_2}{R_1 + R_2}$

In the reciprocal formula, when the sum of the three reciprocals is calculated, don't forget to take the reciprocal of $\dfrac{1}{R_T}$ to get R_T.

Caution! *The $\dfrac{R_1 R_2}{R_1 + R_2}$ formula only applies to two resistors at a time, no more.*

The trick to applying Ohm's Law to circuits is that we can apply Ohm's Law to each individual circuit component but also to the circuit as a whole, using total values for voltage, current, and resistance. A voltage source is said to be a **voltage rise** while the loads are said to drop the voltage, a **voltage drop**. The voltage rises equal the voltage drops. This particular evident in a simple series circuit where the voltage drops at the resistors are a simple sum that must add up to the voltage rise at the battery.

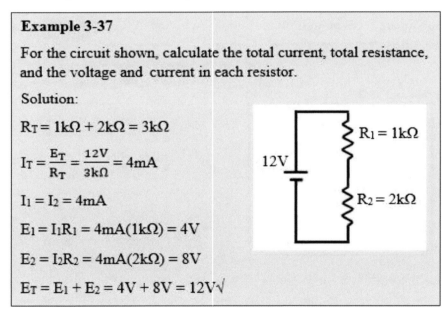

Example 3-37

For the circuit shown, calculate the total current, total resistance, and the voltage and current in each resistor.

Solution:

$R_T = 1k\Omega + 2k\Omega = 3k\Omega$

$I_T = \dfrac{E_T}{R_T} = \dfrac{12V}{3k\Omega} = 4mA$

$I_1 = I_2 = 4mA$

$E_1 = I_1 R_1 = 4mA(1k\Omega) = 4V$

$E_2 = I_2 R_2 = 4mA(2k\Omega) = 8V$

$E_T = E_1 + E_2 = 4V + 8V = 12V \checkmark$

Notice how in milli (m) and kilo (k) cancel when multiplied.

Example 3-38

For the circuit shown, calculate the total current, total resistance, and the voltage and current at each resistor.

Solution:

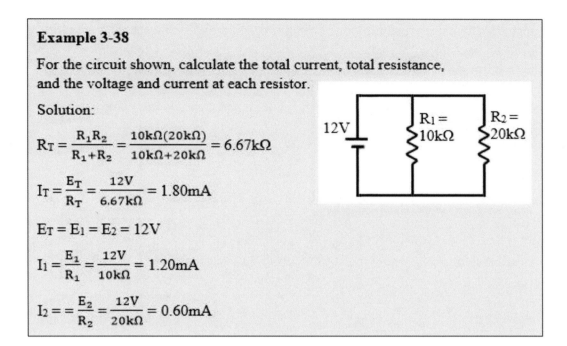

$$R_T = \frac{R_1 R_2}{R_1 + R_2} = \frac{10k\Omega(20k\Omega)}{10k\Omega + 20k\Omega} = 6.67k\Omega$$

$$I_T = \frac{E_T}{R_T} = \frac{12V}{6.67k\Omega} = 1.80mA$$

$$E_T = E_1 = E_2 = 12V$$

$$I_1 = \frac{E_1}{R_1} = \frac{12V}{10k\Omega} = 1.20mA$$

$$I_2 = = \frac{E_2}{R_2} = \frac{12V}{20k\Omega} = 0.60mA$$

Series Limiting Resistor

The purpose of a resistor is to limit current flow and also control the voltage drops on other electronic components. Here we investigate a type of circuit where a resistor must be placed in series so that an electronic component has the appropriate current and voltage. In this case, the value of the voltage source is known. Also known is the voltage and current specifications of a component to be "driven" by this source. We are then to determine the resistor value to be placed in series with the load. The trick is knowing that the voltage across the resistor is the source voltage less the device voltage.

Example 3-39

Calculate the value of the series limiting resistor that will satisfy the specifications of the device.

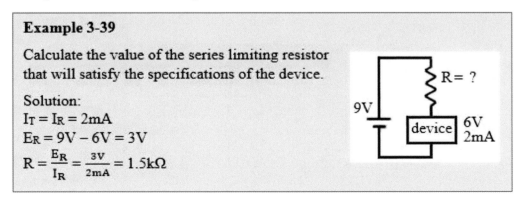

Solution:
$$I_T = I_R = 2mA$$
$$E_R = 9V - 6V = 3V$$
$$R = \frac{E_R}{I_R} = \frac{3V}{2mA} = 1.5k\Omega$$

The Same Resistor Values in Parallel

There are many instances when a great number of resistances of equal value are connected in parallel. For example, say two eight ohm speakers are connected in parallel. The current splits equally between them, cutting the total resistance in half, a four ohm total resistance. This leads us to another formula for calculating total resistance that is useful when there are many resistors of equal value connected in parallel. In a case like this, the total resistance is the value of one resistor divided by the number of resistors.

$$R_T = \frac{R}{N}$$

Example 3-40

Calculate the total resistance when twelve resistors, each 4.7kΩ, are connected in parallel.

Solution:

$$R_T = \frac{R}{N} = \frac{4.7k\Omega}{12} = 392 \ \Omega$$

Resistor Color Code

Resistors are color coded with stripes so that we can read their resistance values without the use of a meter. The resistor is held, or viewed, such that the bands are nearest the end at your left. The first two bands designate two digits of the resistance value. The third band stands for the power-of-ten multiplier. The fourth band indicates the guaranteed accuracy – or tolerance – of the resistance value.

Some resistors have more accuracy and have an extra band. Others have special colors for special tolerances, but shown here are the most common color codes we would encounter.

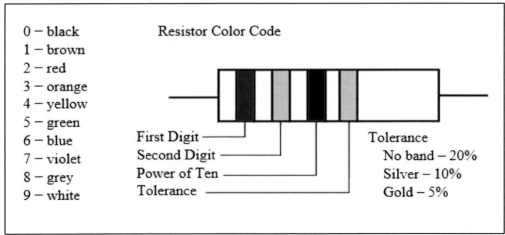

Example 3-41

Interpret the color code of a resistor with the following color code designations: brown-black-red-gold.

    Answer:
    R = 1kΩ with a tolerance of 5%

Example 3-42

Interpret the color code of a resistor with the following color code designations: yellow-violet-orange-silver.

    Answer:
    R = 47kΩ with a tolerance of 10%.

Example 3-43

Interpret the color code of the resistor shown.
    Answer:
    R = 270Ω

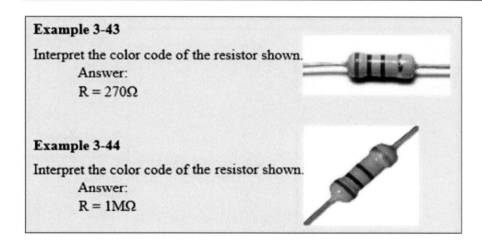

Example 3-44

Interpret the color code of the resistor shown.
    Answer:
    R = 1MΩ

Direct Current and Alternating Current

Electric current may be DC or AC. By DC, we mean **direct current**, which refers to the flowing of charge in one direction. The most common source of direct current is a battery. A battery produces direct current because each terminal of the battery always has the same sign; the positive terminal is always positive and the negative terminal is always negative. Even if the current moves in unsteady pulses, the electrons move in one direction.

In an **alternating current** (AC) circuit, electrons are moved first in one direction and then in the opposite direction, alternating back and forth. This is accomplished by alternating the polarity of the voltage at the generator or other voltage source. Generators produce alternating current.

The alternating voltage "wave" completes one **cycle** when it gradually changes from positive to negative to the point where is begins again. This cycle takes a particular amount of time depending on how quickly the polarity changes. The amount of time the wave takes to complete one cycle is called the **period**, abbreviated with capital letter "T" in formulae. The rate at which the polarity changes is called the **frequency**, abbreviated with the lower case letter "f" in formulae. (Remember how there is an important distinction between the meanings of upper verses lower case letters in formulae.) Frequency is the "speed" of the wave, a rate measured in **cycles per second**, or **cps**. Sometimes this rate is measured in **Hertz**, abbreviated **Hz**. One cycle per second equals one Hertz.

If the period of the wave is a short amount of time, the wave changes from positive to negative quickly. This type of wave has high frequency. If the wave has a longer period of time, the wave is "slower" and has a low frequency. The relationship between the period and frequency of a wave is inverse, or opposite, expressed by the formula $f = 1/T$. Since the period and frequency are reciprocals, the reciprocal key on your calculator can be very handy, simply taking the reciprocal of the frequency to get the period and visa versa.

$$f = \frac{1}{T}$$

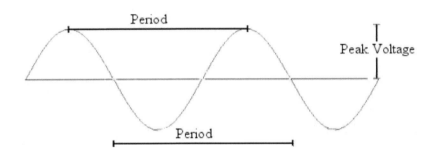

Example 3-45

An AC wave has a period of 20 msec. Calculate the frequency.

Solution:

T = 20 msec

$$f = \frac{1}{T} = \frac{1}{20\text{msec}} = 0.05 \text{ kHz}$$

Note how we get kilo in the answer when dividing my milli. Using these shortcuts saves time and prevents errors, but of course power of ten can be entered into the calculator as well.

Example 3-46

Calculate the period of a 360MHz AC wave. Solution:

$$T = \frac{1}{f} = \frac{1}{360\text{MHz}} = 0.00278\mu\text{sec}$$

Note how we get micro in the answer when dividing my Mega. This is a very fast wave. The greater the frequency, the smaller the time period, an inverse function.

VOLTMETERS & AMMETERS **L9**

Student Name(s)_____

_____

_____

Connecting Voltmeters verses Connecting Ammeters

Objectives

Upon completion of this lab, the student will be able to
- use a digital multi-meter (DMM) to measure DC current and voltage.
- describe what is meant by a series-connection.
- describe what is meant by a parallel-connection.
- explain how voltmeters and ammeters are connected differently in a circuit.

Main Ideas

- Current (amperage) measurements are made by connecting a meter within the electrical path of an electrical circuit, a series connection.
- Voltage measurements are made by connecting the meter across two selected points in an electric circuit, a parallel connection.

Discussion

In this lab you'll learn to use digital multi-meters (DMM) to measure both current in amperes (or milliamps) and voltage in volts (or millivolts).

There is only one path for the current in a series circuit while in a parallel circuit there is at least one point in the circuit where the current can split. Ammeters are connected in series. Voltmeters are connected in parallel.

Ammeters and voltmeters are connected differently in an electric circuit. Voltmeters are simpler because the meter leads are simply placed across (parallel) any component across which the voltage is to be measures. But ammeters must be connected in series, that is, it must be part of the circuit, with current flowing through the meter.

Figures at right show block diagrams depicting the two ways these meters are connected. Note how in the voltmeter is connected in parallel across the load while in the ammeter is in series as part of the circuit.

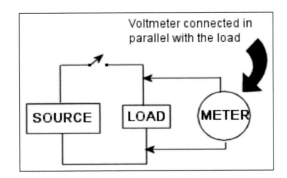

Voltmeter connected in parallel with the load

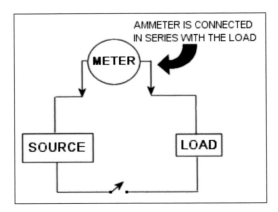

AMMETER IS CONNECTED IN SERIES WITH THE LOAD

Apparatus

Two digital multi-meters
Banana leads
DC power supply or Battery Eliminator
Banana patch bay
Knife switch (spst), optional
6-volt lamps

At right, component patch bay containing switch and lamps together with power supply, banana wires, and meters connected for load voltage and current.

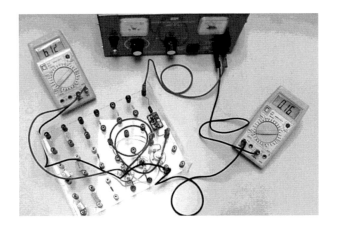

Procedure

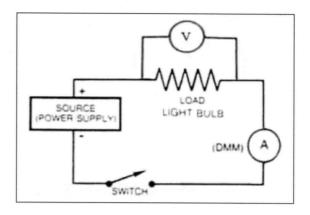

1. Connect the apparatus as shown at right. Be sure the power supply is off, the switch is open, and the DMMs are set to off. Use one bulb in the patch bay. The power supply leads are connected to the DC voltage terminals.
2. Designate one meter as the voltmeter. Set its function to volts. Connect this meter across the load in parallel. Designate the other meter as the ammeter. Connect this meter in series with the load.
3. Set the ammeter function to DC current. This is sometimes done differently on different brands of DMMs, but in general the DC current function must be selected and also a range of up to 10 amps. Make sure the meter leads are connected to the correct DC amp jacks on the meter.
4. Select the voltmeter function to DC volts and select the appropriate range setting on the meter, say 20 VDC. Make sure the leads are connected to the correct jacks on the meter.
5. Turn the meters on.
6. Turn the power supply voltage control to its minimum setting (completely counter clockwise). Turn the power supply on by selecting DC volts.
7. Close the knife switch. Turn up the voltage on the power supply voltage control. Keeping an eye on the voltmeter, *be careful not to exceed 6 volts*, the rated voltage of the patch bay bulb. Set the voltage to 2 volts. The bulb should light up and an amperage reading should appear on the ammeter. Note the current reading in the data table.
8. Increase the voltage to 4 and 6 volts, recording the current for each in the data table.

Data

Voltage Setting	Current
2 volts	
4 volts	
6 volts	

Questions

1. If all the components of a circuit remain the same, but the source voltage is increased, what happens to the amount of current flowing in a circuit? What parts of this lab activity support your answer?

2. What is precision and accuracy of each of the meters used in this lab?

3. Explain in your own words what is meant by a series-connected meter.

4. Explain in your own words what is meant by a parallel-connected meter.

Count Alessandro Giuseppe Antonio Anastasio Volta (1745 – 1827)

Count Volta was an Italian physicist who revolutionized science and technology by his invention of the electric battery. This invention was based on his many years of study of electrical phenomena, after he had won a reputation in the field. It inspired research in a wide range of scientific fields, from chemistry to physics to medicine, and laid the foundation for the age of electronics.

The unit of electrical potential, the volt, is named after him.

LABORATORY ACTIVITY

OHM'S LAW L10

Student Name(s) ____________________

____

____

Ohm's Law: Resistors in Series and Parallel

Objectives

Upon finishing this lab, the student will…

- distinguish between series and parallel circuits.
- define Ohm's Law.
- correctly connect a multi-meter in a circuit to measure voltage and current.
- use Ohm's Law to calculate resistance.
- experimentally verify Ohm's Law.
- calculate percent error.

Discussion

Ohm's Law Formula

Ohm's Law is a formula that relates voltage, current, and resistance in electrical circuits. The voltage E (or sometimes V) is always a voltage *difference* (often written ΔE) between two points in a circuit. The current I is the rate of the flow of charge between these points. The resistance R is the "electrical friction" between the points that slows down the charge flow rate and so reduces the current flow. The relationship between these three quantities is expressed as Ohm's Law

$$E = IR$$

where E is the voltage in volts (V), I is the current in amperes (A), and R is the resistance in ohms (Ω, capital Greek letter omega).

Voltage

It is important to remember that E is the *voltage difference* between two particular points in a circuit. It is the "mover" quantity that causes electrons to flow. A voltmeter measures this difference when its two terminals are connected in parallel across the two points.

Resistance

Any electrical load will resist current flow. Without a resistive load there is a short circuit. The resistance of the load is measured in ohms (Ω). Most multi-meters can be used to measure this resistance, but this must be done with the device out-of-circuit. When a circuit component, say a resistor, is removed from a circuit to measure its resistance, this measurement is called the *cold resistance.* When the resistance of the component is calculated using Ohm's Law, this indirect measurement is called the *hot resistance* (because resistors warm up when under load).

Current

The current (I) through the resistor (R) can be measured using a multi-meter. Remember that in order to measure current, the ammeter must be connected in series, or "in line", with the load. With only one path for the current to flow, in series, the current through the resistor has to be the same as the current through the ammeter.

Shown at right is a portion of a circuit, just a resistor. The schematic symbol for a resistor is a zig-zag. Note also that the voltmeter is connected across (in parallel) the resistor while the ammeter is connected in series. Only one wire lead is needed to connect an ammeter into a circuit; one of the leads is already in the circuit. The circuit must be opened, broken, somewhere and ammeter is placed in this break to complete the circuit.

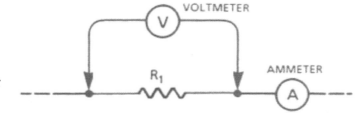

Series verses Parallel Circuits

Circuit **a** below is called a "single loop" or "series" circuit because there is only one path for current to take as is flows from one terminal of the battery to the other. This is a series circuit. In a series circuit, the current is the same in each resistor while the voltage is shared.

Series:

$$I_T = I_1 = I_2 = I_3$$

$$E_T = E_1 + E_2 + E_3$$

a.

By contrast, in circuit **b** below, there are three basic loops. The current splits, with the three resistors sharing the total current coming from the voltage source. This is a parallel circuit. In a parallel circuit, the voltage is the same on each resistor, with each their own connection across the battery, but the current is shared.

Parallel:

$$E_T = E_1 = E_2 = E_3$$

$$I_T = I_1 + I_2 + I_3$$

b.

Calculating Total Resistance

The voltage source acts as the "mover", and "pushes" electrons through the resistors. The total resistance the voltage source "feels" not only depends on the individual values of each resistor, but also by how they are connected. The voltage source "feels" only the total resistance R_T of the circuit. Total resistance is calculated depending on how the resistors are connected.

In *series circuits*, the total resistance is simply the *sum total* of all the value of the resistors:

$$R_T = R_1 + R_2 + R_3$$

In parallel circuits, the resistance is considerably less because the current has more paths it can take. With current increasing with each parallel branch, it is the reciprocals of the resistances that add. The total resistance of a parallel circuit can be calculated using the formula

$$\frac{1}{R_T} = \frac{1}{R_1} + \frac{1}{R_2} + \frac{1}{R_3}$$

Resistor Color Code

Earlier we studied the resistor color codes. Resistors are color coded with stripes so that we can read their resistance values without the use of a meter and when they are too small to print numbers on. Shown below is the color code table. The first two bands designate two digits of the resistance value. The third band stands for the power-of-ten multiplier. The fourth band indicates the guaranteed accuracy – or tolerance – of the resistance value.

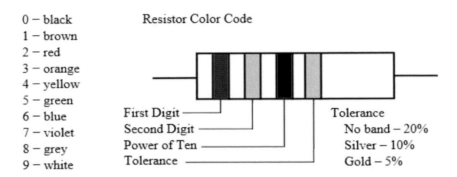

Resistor Color Code

0 – black
1 – brown
2 – red
3 – orange
4 – yellow
5 – green
6 – blue
7 – violet
8 – grey
9 – white

First Digit
Second Digit
Power of Ten
Tolerance

Tolerance
No band – 20%
Silver – 10%
Gold – 5%

Calculating Percent Error

Percent error is useful when comparing how far off the predicted voltage is to the actual measured voltage. This is calculated:

$$\% \text{ error} = \frac{|\text{ predicted voltage} - \text{measured voltage }|}{\text{measured voltage}}$$

Equipment

Circuits Trainer
DC Power Supply or
    Battery Eliminator
Banana connectors
Two digital multi-meters
Resistors:
    $100 \, \Omega$
    $220 \, \Omega$
    $560 \, \Omega$
    $1.0 \, k\Omega$
    $3.0 \, k\Omega$
    $4.3 \, k\Omega$

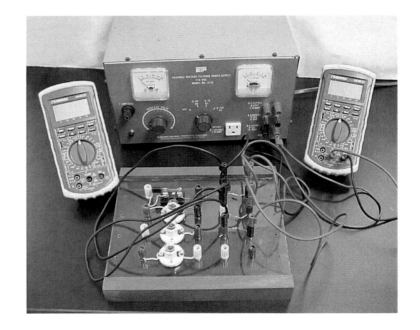

Note: You have the option of using actual ohmmeter measurements of the resistors in your calculations rather than the color code values. This eliminates error due to resistor tolerances.

Procedures

Part A: One Resistor

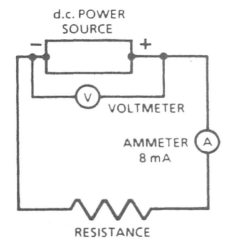

1. Set up the DC single-resistor circuit shown at right. Use a 1.0 kΩ (kilo means times 1000, so this is a 1000 ohm resistor).

2. Use Ohm's Law to calculate the amount of voltage required to cause 8mA (milli means one thousandth) of current to flow in the circuit. Record this calculated value in the Data Table A.

3. With the DC power supply initially set for 0 volts output, gradually increase the voltage until the current reading on the ammeter is 8mA.

4. Read the voltage indicated on the voltmeter and record this value in the Data Table A.

Part B: Three Resistors in Series

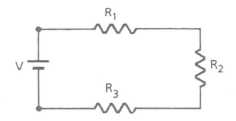

1. Set up the series circuit shown at right. Use resistances

   $R_1 = 100\,\Omega$

   $R_2 = 220\,\Omega$

   $R_3 = 560\,\Omega$

2. Use the series resistance formula to calculate total resistance R_T. Record this value in Data Table B.

3. Using the above calculated R_T and Ohm's Law, calculate the amount of voltage required to cause 8mA of current to flow in the circuit. Record this value in Data Table B.

4. With the DC power supply initially set at 0 volts, gradually increase the voltage until the current reading on the ammeter is 8mA.

5. Read the voltage indicated on the voltmeter and record it in Data Table B.

Part C: Three Resistors in Parallel

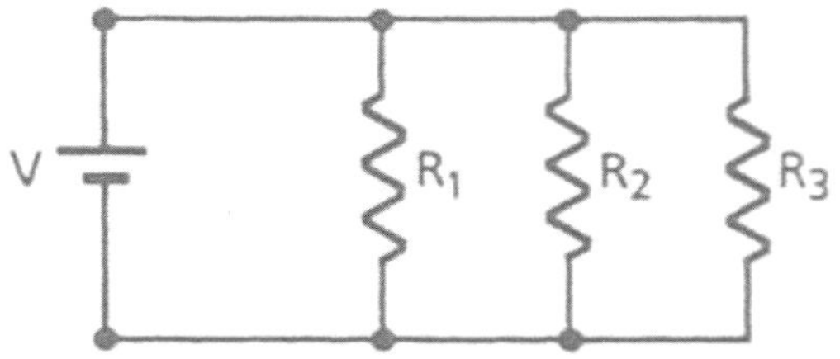

1. Set up the parallel circuit as shown at right. Again use
   $R_1 = 1.0\,k\Omega$
   $R_2 = 3.0k\Omega$
   $R_3 = 4.3k\Omega$

2. Using the parallel resistance formula for R_T, calculate the total resistance of the circuit. Record this value in Data Table C.

3. Using the above calculated R_T and Ohm's Law, calculate the amount of voltage required to cause 8mA of current to flow in the circuit. Record this value in Data Table C.

4. With the DC power supply initially set at o volts, gradually increase the voltage until the current reading on the ammeter is 8mA.

5. Read the voltage indicated on the voltmeter and record it in Data Table C.

Data Tables

Part A	
Total Resistance	Ω
Predicted (calculated) Voltage	V
Measured Voltage	V
% Error	%

Part B	
Total Resistance	Ω
Predicted (calculated) Voltage	V
Measured Voltage	V
% Error	%

Part C	
Total Resistance	Ω
Predicted (calculated) Voltage	V
Measured Voltage	V
% Error	%

Questions

1. Explain how a voltmeter and an ammeter can be used to determine the total resistance of an electric circuit.

2. Explain the basic difference between a series circuit and a parallel circuit.

3. Explain in words Ohm's Law.

4. If an $8\,k\Omega$ resistor and a $12\,k\Omega$ resistor are in parallel across a 12 volt battery, how much current will flow?

5. If a $1\,k\Omega$, a $3\,k\Omega$, and a $4\,k\Omega$ resistor are in series across a 12 volt battery, how much current will flow?

6. Why is there more total resistance when the resistors are connected in series and less total resistance when the resistors are connected in parallel?

Georg Simon Ohm (1787 - 1854)

Although Georg Ohm discovered one of the most fundamental laws of electricity, he was virtually ignored for most of his life by scientists in his own country, Germany.

While a child, Ohm's ambition was to become a scientist and to work at one of the great German Universities. Ohms' father was a mechanical engineer and taught him basic practical skills that later proved useful.

Ohm's main interest was current electricity, recently advanced by Alessandro Volta's invention of the battery. He made his own metal wire, producing a range of thicknesses and lengths of remarkably consistent quality.

In 1827, he was able to show from his experiments that there was a simple relationship between resistance, current and voltage. Ohm was afraid that his purely experimental basis of his work would undermine the importance of his discovery. He tried to state his law theoretically and his rambling mathematical proofs made him an object of ridicule. He was criticized so strongly that he was forced to resign his teaching post.

In the years that followed, Ohm lived in poverty and isolation. In 1842, the Royal Society in London recognized the significance of his discovery and admitted him as a member. In 1849, just five years before his death, Ohm's lifelong dream was realized when he was given a professorship at the University of Munich.

The standard unit of electrical resistance, the Ohm (Ω), is named after him.

Measuring Frequency with an Oscilloscope

Main Ideas

- Function generators can be used to develop different types of waveforms or electrical signals. These can be displayed on an oscilloscope.
- Oscilloscopes can be used to measure the amplitude and period of an electrical signal (waveform).
- The frequency of a repeating electrical signal is equal to the number of cycles (repeats) of the signal per second.
- Oscilloscopes can be used to determine the frequency of a signal by measuring the period.
- The frequency (f) and period (T) of a repeating electrical signal are inverses of one another, that is $T = 1/f$ and $f = 1/T$.
- The peak-to-peak amplitude (voltage) of an electrical wave is twice the peak amplitude.

Discussion

In DC circuits we learned that the voltage stays the same and does not change, with current flowing steadily in one direction from positive to negative. In this lab we will study how the voltage repeatedly changes polarity. This *rate* of the alternating polarity is called alternating voltage. Of course, when the voltage alternates, the current alternates correspondingly, hence alternating current, or AC. In this lab we will display this alternating voltage on a device called an *oscilloscope*, a display from which we will determine amplitude, period, and frequency.

Frequency refers to the number of events that happen in a unit of time. In electrical circuits, the event that is counted is sometimes called a "voltage cycle". A voltage cycle is merely a voltage signal that changes from positive to negative in a given period of time over and over again. These patterns are also called "waveforms".

When graphically depicting (drawing) these waveforms, the vertical axis represents the voltage while the horizontal axis represents the time.

When a single cycle of a waveform occurs once in one second, the frequency is one cycle per second (or 1 cps). One cycle per second is called "one hertz". If the pattern occurs twice in one second, the frequency is 2 cps or two hertz. In various technologies, frequencies can be less than one hertz or frequencies well over millions of hertz. **Hertz** is the measure of electrical frequency.

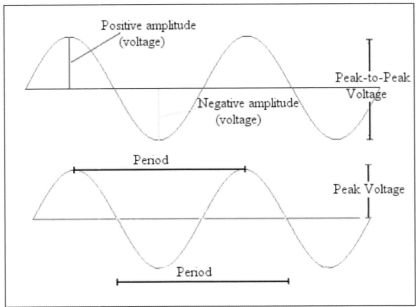

The above figure diagrams the names of the parts of a waveform. The period (T) of a waveform is the time it takes to complete one cycle. Note how the position of the beginning of the cycle doesn't matter as long as the end of the cycle is the point where it begins to repeat. The amplitude of electrical waves is the maximum voltage where the peak voltage is half the peak-to-peak (p-p) voltage. Note that the peak-to-peak voltage is the absolute sum of the positive and negative parts of the voltage cycle.

Frequency is not directly measured from the oscilloscope, but calculated by measuring the period.

In this lab you will operate an oscilloscope and a function generator. The function generator simply produces the waveforms to be measured. The oscilloscope will be used to display these waveforms.

A **function generator** produces different types of voltage waveforms. Different types of waveforms include the sine wave, square wave, triangle wave, and saw-tooth wave. Functions generators have four principle controls: the wave type selector, an amplitude control (voltage), a frequency range, and a fine variable frequency control.

The **oscilloscope** produces a visual graph of a waveform from which period and amplitude can be measured from the graphical grid on the screen face. The oscilloscope graphically depicts how the voltage changes with respect to time, with voltage on the vertical axis and time on the horizontal axis. The line produced on the oscilloscope screen is called a "trace". Although the controls might at first look a little complicated, it has four basic categories of control knobs and switches: Trace-quality, vertical scale and position, horizontal scale and position, and display stability.

Equipment

Oscilloscope, shown here the old CRT type, but your instructor might provide a newer type
Function generator
Connecting cables or scope probe

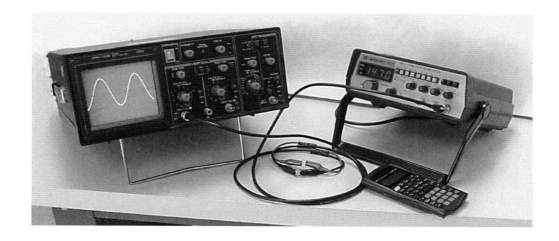

The lab setup. Note that the function generator is connected directly to the input of one of the oscilloscope channels.

Procedure

1. Connect the output of the function generator to one of the input channels (more than one signal can be displayed at a time) of the oscilloscope.

2. Turn both instruments on, letting them both warm up for a moment or two.

3. Select the sine wave function on the function generator.

4. On the function generator, turn the amplitude control to the maximum level.

5. Set the frequency control by pressing any one of the center range buttons available on the function generator.

6. On the oscilloscope, adjust the focus, intensity, vertical scale, and horizontal scale such that a steady trace of the waveform is displayed on the screen. (Toying with the knobs is recommended here, experimenting with their functions. You cannot damage anything. Ask the instructor for help if needed.)

7. Amplitude: Once a steady, complete wave is displayed, on the oscilloscope, read the volt-per-division setting at the voltage control. Record this volts- per-division setting in the data table.

8. On the oscilloscope, count the number of vertical divisions representing p-p voltage from the screen grid, recording this in the data table.

9. Period: From the oscilloscope time control, read the horizontal time-per-division setting and record this in the data table.

10. On the oscilloscope screen, count the number of horizontal divisions representing one cycle from the screen grid, recording this in the data table.

11. Calculate the peak-to-peak voltage, period, and frequency of the displayed wave using your measurements.

Data Table

Volts per division control setting _____

Number of vertical divisions _____

Time per division control setting _____

Number of horizontal screen divisions _____

Calculations

Peak-to-peak voltage:
    E_{p-p} = volts/division times the number of vertical divisions = _____

Period:
    T = time/division times the number of horizontal divisions = _____

Frequency:
    $f = \dfrac{1}{T} =$ _____

QUESTIONS

1. Explain in your own words what *amplitude* and *period* are with respect to electrical waveforms.

2. What is meant by the *frequency* of electrical waveforms?

3. What is the relationship between the *period* of a wave and its *frequency*?

4. Explain how p-p voltage and frequency of an electrical waveform are determined using an oscilloscope.

1. R_1 is color coded brown-black-red and R_2 is color coded orange-black-red. For E = 12 volts, calculate

   (a) the total resistance R_T,

   (b) the total current flow I_T, and

   (c) the current and voltage at each individual resistor.

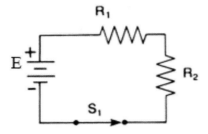

2. R_1 = 60 kΩ, R_2 = 100 kΩ, and R_3 = 150 kΩ. For E = 12 volts, calculate

   (a) the total resistance R_T,

   (b) the total current flow I_T, and

   (c) the current and voltage at each individual resistor.

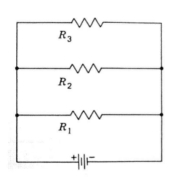

3. A 6 volt circuit supplies current to a single 1.5 volt device. The current through the device is must be limited to 2 mA. Find the value of a series limiting resistor needed to limit the current to 2 mA.

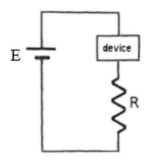

4. Each resistor is $120\,k\Omega$. Total current is 3.2 mA. Calculate the applied voltage.

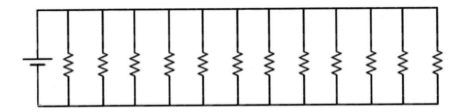

5. Given $E_1 = 36V$, $I_1 = 0.4A$, $E_2 = 24V$, $E_3 = 16V$ find I_2, I_3, I_T, R_1, R_2, R_3, R_T, and E_T.

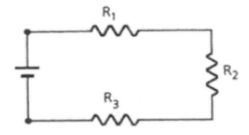

6. Given $R_3 = 100\Omega$, $I_1 = 200mA$, $I_2 = 300mA$, $I_3 = 400mA$ find E_T, R_T, R_1, R_2, and I_T.

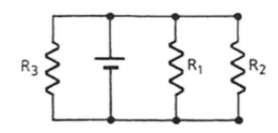

7. The period of an AC wave is 2.5μsec. Calculate the frequency.

8. Find the peak-to-peak voltage and frequency of the square wave given the oscilloscope settings shown.

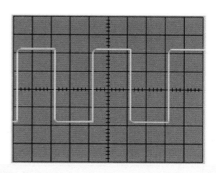

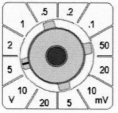 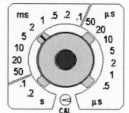

9. Find the peak-to-peak voltage and frequency of the sine wave given the oscilloscope settings shown.

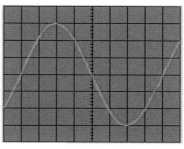

 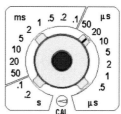

Answers:
1. $R_T = 4k\Omega, I_T = I_1 = I_2 = 3mA, E_1 = 3V, E_2 = 9V$
2. $R_T = 30k\Omega, E_1 = E_2 = E_3 = 12V, I_T = 0.4mA, I_1 = 0.20mA, I_2 = 0.12mA, I_3 = 0.08mA$
3. $R = 2.25k\Omega$
4. $E = 32V$
5. $E_T = 76V, I_T = I_{1,2,3} = 0.4A, R_1 = 90\Omega, R_2 = 60\Omega, R_3 = 40\Omega, R_T = 190\Omega$
6. $I_T = 900mA, E_T = E_{1,2,3} = 40V, R_2 = 200\Omega, R_3 = 133\Omega, R_T = 44.4\Omega.$
7. 400kHz
8. $20.0\ V_{p-p}$, 125 Hz
9. $60\ V_{p-p}$, 6250 Hz

Our unifying principle states that force, torque, fluid pressure, voltage and temperature difference all are movers in their respective energy systems. Temperature difference moves heat in thermal systems.

Temperature (T)

When a substance warms up and gets hotter, the particles in the substance vibrate and get more energetic. **Temperature** is a measure of this activity, a measure of hotness or coldness. It might seem strange, but the temperature is not the measure of heat in the substance, but how the substance reacts to this heat. Different substances react differently to heat, a quantity called "specific heat", which was discussed above. Temperature and heat are interrelated, but different, quantities.

There are a few different standards of measurement for temperature, but here we will study two, the **Fahrenheit** (°F) scale and the **Celsius** (°C) scale. Sometimes the Celsius scaled is called the Centigrade scale, in French meaning "graduated in 100^{ths}". Sometimes it is necessary to convert between °F and °C. This is done with the use of the following formulae. Actually it's one formula, manipulated to isolate either F or C. To keep things simple, F represents the temperature (T) in °F and C represents the temperature (T) in °C.

$$C = \frac{5}{9}(F-32) \qquad F = \frac{9}{5}C+32$$

One or the other formula is selected depending on what is known and what is to be calculated. Note that although the formulae look somewhat similar, they contain different orders of operations. The formula to calculate C, on the left, has parentheses and subtraction. This subtraction must be done first. The formula to calculate F, on the right, has no parentheses so the multiplication must be done first, then the addition of 32.

Temperature Difference (ΔT)

Temperature (T) and temperature difference (ΔT) are different quantities. Note how temperature difference is represented with ΔT. The Greek letter Δ (delta) does not represent a separate quantity. We do not substitute numbers in for Δ. This is a letter that is placed in front of another letter, a literal, to mean "change in" or "difference". Any change or difference is calculated by subtracting.

In this case the subscripts are "f" for final and "i" for initial. The change in temperature is the difference between the object's final temperature and what the temperature initially was.

$$\Delta T = T_f - T_i$$

The units of measurement for ΔT is different than T. For temperature (T), the units are written °F and °C. For temperature difference (ΔT) the units are written F° and C°. Note that the degree symbol precedes the F or C for temperature (T) but follows the F or C for ΔT. This might seem like a trivial thing, but think how useful it can be. This eliminates ambiguity between the two quantities when values are given. For example, by writing the units differently, there is no confusion between 54°F, a temperature, and 54F°, a temperature change or difference. This seemingly insignificant way in writing the units can be very important, but is often misconstrued, even in physics text books.

There's a common mistake when converting ΔT between F° and C°. Do not use the temperature conversion formulae above. ΔT conversions are done differently. It's even easier. How this works may best be explained with the use of a liquid-in-glass thermometer. These types of thermometers have the air removed, eliminating atmospheric pressure. This allows the liquid inside the hair thin shaft to expand and contract freely as the temperature changes.

Take a look at the thermometer at right. There is a 100C° change when the temperature changes from 0°C to 100°C. The corresponding temperature change in Fahrenheit is 180F°, going from 32°F to 212°F. So a temperature change of 100C° is equal to 180F°.

ΔT
100C° = 180F°
1C° = 1.8F°

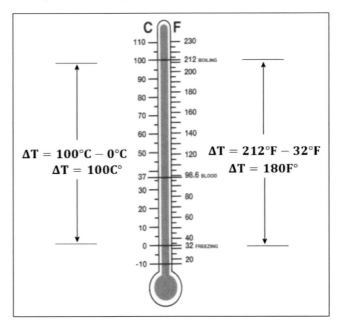

The formulae that are used to convert temperatures (T) will not work to convert ΔT. For example, if 100C° is substituted into $F = \frac{9}{5} C + 32$ the result would be 212, not 180! Do not use those formulae to convert temperature changes.

Since 1C° = 1.8F°, temperature difference is converted by multiplying or dividing by 1.8.

Examples 3-47

   a. Convert T = 50^{0}F to ^{0}C.

     Solution:

$$C = \frac{5}{9}(F-32) = \frac{5}{9}(50-32) = 10^oC$$

   b. Convert T = $-$ 15^{0}C to ^{0}F

     Solution:

$$F = \frac{9}{5}C+32 = \frac{9}{5}(-15) + 32 = 5^oF$$

   c. Convert ΔT = 70F^o to C^o

     Solution:

     $1C^\circ = 1.8F^\circ$

$$\Delta T = 70\cancel{F^o}\left(\frac{1C^0}{1.8\cancel{F^0}}\right) = 38.9C^0$$

   d. Convert ΔT = 20C^o to F^o

     Solution:

     $1C^\circ = 1.8F^\circ$

$$\Delta T = 20\cancel{C^o}\left(\frac{1.8F^0}{1\cancel{C^0}}\right) = 36F^0$$

   e. Convert T = 20^{0}C to ^{0}F

     Solution:

$$F = \frac{9}{5}C+32 = \frac{9}{5}(20) + 32 = 68^oF$$

Student Names(s) _____

_____

_____

Objectives:

When you've finished this lab, you should be able to…
- briefly explain how a thermocouple works.
- set up a thermocouple circuit using a voltmeter, thermocouple wire, and ice.
- use a thermocouple calibration table to find the temperature given the thermocouple voltage.
- measure temperature using a thermocouple.

Main Ideas:

- A thermocouple is a junction (wires twisted or fused together) of dissimilar metals that produces a small voltage proportional to its temperature.
- The *type* of thermocouple depends on the kinds of metals used to form junctions.
- Reference junctions, usually $0°C$ or $32°F$, combine with e test junction for proper measurement.

Discussion:

A thermocouple is a junction (wires twisted or fused together) of dissimilar metals that produces a small voltage proportional to its temperature. It's simple, rugged, and easy to use. Other advantages are that a thermocouple can be used in places and for temperature ranges where ordinary mercury or alcohol thermometers are impractical or cannot be used at all. Also, since a voltage is produced at the junction, thermocouples are compatible with other electrical control systems, including computers. Many digital multi-meters now have built-in thermocouples. (You're welcomed to use one during this lab for comparison with your ice-junction thermocouple.) These benefits make the thermocouple the most widely used temperature-measuring device used in industry.

In use, the *reference junction* of the thermocouple is kept at a known temperature in order for the calibration tables to work. This temperature is usually the freezing point of water, $0°C$ or $32°F$. This reference junction temperature is maintained by placing the junction in ice-water. The other junction, the *measurement junction*, is then brought into contact with the substance whose temperature is to be measured. The voltage reading made by a voltmeter is then proportional to the difference between the voltages generated by the two junctions. Knowing the voltage, the temperature can be determined with a calibration table.

To learn how to read the thermocouple table, obtain the *Omega Temperature Handbook*. When reading the thermocouple calibration table, one must first look up the appropriate type of thermocouple used, here type-K, this type usually color coded yellow. The table has a vertical column of temperatures and also a horizontal row of temperatures. This is done to save space. First locate the nearest value voltage in the table, then add the temperature at the left column with the temperature in the bottom horizontal row. Check these examples in the handbook. For Celsius yellow-red type-K:

1.500 mV is 30 + 7 = 37°C   and   6.300 mV is 150 + 4 = 154°C

Equipment:

Several different *types* of thermocouples are made. The one you'll use in this lab is a Type-K. Manufacturers of these types of wire provide calibration table free with their catalogues. Here you will need the *Omega Temperature Handbook* available in the laboratory.

Omega Temperature Handbook
   (or type K table in in Reference Tables of this text)
Thermometer thermocouple meter
Type-K thermocouple wire (yellow-red)
Hot Plate
Two beakers (or other containers)
Ice
Digital multi-meter
Banana leads
Patch bay

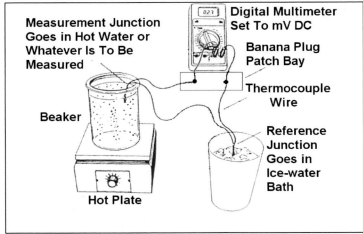

Procedure:

1. Start some water boiling on the hot plate.
2. Fill the second beaker with water and crushed ice.
3. Construct a type-k thermocouple by twisting the ends of three strands together such that there are two junctions, one the test junction, the other the ice reference junction.
4. Connect the free ends of the thermocouple wires to the patch bay by inserting the wire into the lug opening and tightening the lugs.
5. Using two banana leads, connect the voltmeter to the patch bay. The milli-voltmeter is to measure the sum of the voltages of the reference and measurement junctions.
6. Place either one of the junctions in the ice water. This is the reference junction. Let the other junction lie freely on the table. This junction should be at room temperature. The voltage reading on the multi-meter should be a positive reading; if not, reverse the banana plugs on the meter. Record the meter reading under "Thermocouple Voltage" in the Data Table opposite "Room Air".
7. With the reference junction in the ice water, place the measurement junction in the boiling water. The water should be close to $100°C$ or $212°F$, but this temperature can vary with altitude and weather conditions. Measure this temperature with the thermocouple meter, recording the voltage reading in the Data Table opposite "Boiling Water".
8. Refer to the type-K calibration table in the Omega Temperature book (there is a truncated version of the table at the back of this book in the Reference Tables). For the two voltages listed in the Data Table, use the calibration table to find the corresponding temperature. Record each of these temperatures in the Data Table.
9. Measure the room temperature using the thermocouple meter, recording this in the Data Table.

Data:

	Thermocouple Voltage	Temperature from Calibration Table	Temperature using Multi-Meter	Percent Error
Room Air				
Boiling Water				

Percent Error:

The actual temperature of the room air is known. Also, the temperature of boiling water is known to be $100°C$. Compare these actual temperature values to the experimental by calculating the percent error of your results using the formula below. Record the percent error in the Data Table.

$$\% \text{ error} = \frac{|\text{experimental} - \text{actual}|}{\text{actual}}$$

Thermocouples Wrap-Up:

1. Name two major reasons why thermocouple thermometers are far superior to liquid-in-glass thermometers.

2. Explain in your own words how a thermocouple works.

3. What is meant by the *type* of thermocouple?

4. Explain how to read a thermocouple calibration table.

5. Explain why the ice junction is necessary.

3.5 STUDENT EXERCISES   TEMPERATURE & TEMPERATURE DIFFERENCE

1. Convert T = 67°F to °C.

2. Convert T = 26°C to °F.

3. Convert T = 0°C to °F.

4. Convert ΔT = 100C° to F°.

5. Convert ΔT = 160F° to C°.

6. The temperature of a food product is flash-frozen from room temperature, 68°F, to −20°F. Calculate the temperature change.

7. Explain how a thermocouple thermometer works.

8. Why is it necessary to use a cold junction compensator for a thermocouple to work?

9. Use the Type-K thermocouple table in the reference tables at the back of this book to look up the temperature that corresponds to a 0.67mV thermocouple multi-meter reading.

10. The Type K thermocouple reading is 2.85mV. What is the corresponding thermocouple temperature?

Answers: 1. 19.4°C **2.** 78.8°F **3.** 32.0°F **4.** 180F° **5.** 88.9C° **6.** −88.0F° **7.** *The junction of two dissimilar metals produces a small voltage proportional to the junction temperature above ambient. The temperature is then found by looking up this voltage in a thermocouple table.* **8.** *Without the compensator, the voltage is zero at ambient temperature (whatever the room temp is). By using a cold junction connected in series with the test junction, the cold junction voltage is added to the test junction, thereby establishing a common reference that will make the thermocouple tables work at any ambient temperature.* **9.** 62°F **10.** 158°F

Chapter 4   Work & Energy

The term "work" can have many meanings depending on how it is applied in everyday conversations. But this word has a very specific technical definition. In a scientific sense, work is done only when movement or some kind of change has occurred in a system. It takes "energy" to do work. "Energy" and "work" are closely related. They have the same units of measurement. **Energy** is the ability to do **work**, often thought as stored, while **work** is a measure of what is actually done, or energy used. For example, a battery has stored energy and the ability to do work, but no work is done until charge moves between the battery terminals, maybe starting a car. Another example might be energy stored in the form of compressed air which then does work by displacing the air and driving a pneumatic tool.

In earlier sections we learned about the displacement quantities (the measure of change in a system) and the mover quantities (the quantities that cause the changes) in each energy system. Work is the product of these two quantities.

Unifying Principle for Work

Work = Mover x Displacement

Although this unifying principle does not apply to thermal systems, an exception we will explore later, the unifying formula still neatly applies to the remaining four energy systems. As long as we understand each mover and displacement quantity of each energy system, we can apply a powerful unifying concept to four energy systems.

Energy System	Mover	Displacement	Work = Mover x Displacement
Translational Mech	Force (F)	Distance (d)	$W = Fd$
Rotational Mech	Torque (τ)	Angle (θ)	$W = \tau\theta$
Fluid	Pressure (p)	Volume (V)	$W = pV$
Electrical	Voltage (E)	Charge (q)	$W = Eq$

4.1   Work in Translational Mechanical Systems

Translational mechanical work is done when a force moves an object a distance. Here we see that force is multiplied to distance. As a result, the units of measurement for work and energy is foot-pounds (ft-lb) in the English system and Newton-meters (N-m) in SI. The Newton-meter has been named after yet another great scientist, Joule.

$$W = Fd$$

$$1 \text{ N-m} = 1 \text{ J}$$

Much of our study of translation mechanical work will be in the lifting of weights, where the force applied must be at least equal to the weight of the object being lifted and the distance is the height it is lifted.

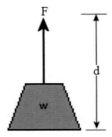

Example 4-1

Calculate the work required to lift a 500 pound object to a height of 60 feet.

Solution:
$$W = Fd = 500 \text{ lb } (60 \text{ ft}) = 30{,}000 \text{ lb-ft}$$

Below is a metric version of the same scenario.

Example 4-2

Calculate the work required to lift a 10 kN object to a height of 30 meters.

Solution:
$$W = Fd = 10\text{kN } (30 \text{ m}) = 300 \text{ kN-m} = 300 \text{ kJ}$$

Remember that 1 N-m = 1 Joule. It is important here to understand the difference between units of torque and units of work, which appear the same. When we studied torque, we learned that torque is calculated by multiplying the force to the lever arm. In a sense the units are indeed the same as work where we multiply a force times a distance. But in rotational systems, the force and the lever arm are *perpendicular*, while for work the force and distance are on the *same line of action*. The relationship between force and "distance" are very different in the two cases. Although the units appear to be the same, they are a measure of two completely different quantities. To avoid confusion, torque units are ft-lbs and N-m. When it's work and energy the units are lb-ft, written backward in the English system, and Joules in SI. This convention of writing and saying the units distinguishes between the two

120 ft-lb	(Torque)
120 lb-ft	(Work & Energy)
120 N-m	(Torque)
120 J	(Work & Energy)

quantities with similar units. See how when given just the measurement and units, the writing convention make clear which is which.

Example 4-3

How much work is required to raise 100 kg to a height of 20 meters?

Solution:

$$w = mg = 100kg\,(9.81m/s^2) = 981N$$
$$W = Fd = 981N(20m) = 19620\,J = 19.6\,kJ$$

Note how the weight in Newtons must be calculated first in order to get units of work in Joules.

Example 4-4

How much weight can be lifted to a height of 60 feet when 96,000 lb-ft of energy is available?

Solution:

$$F = W/d = 96,000\text{lb-ft}/60\text{ft} = 1600\,lb$$

Potential and Kinetic Energy

When an object it lifted to a particular height, the work done is W = Fd. At this height the object is said to have **potential energy** because it could fall, with the Earth's gravity pulling it back down. A raised object is said to have potential energy much like a compressed air tank or charged battery has potential energy, the ability to do work. The formula for the potential energy of a raised object is the same as W = Fd except that instead of "W" for work, "PE" for energy is used, instead of "F" for force, "w" for weight is used, and instead of "d" for distance, "h" for height is used. Also, we must remember to calculate the weight if given the mass, as shown in example 4-3.

The units for potential energy are the same as work, lb-ft in the English system and Joules in SI.

Potential Energy of a Raised Object

PE = wh   (when weight is given)

PE = mgh   (when mass is given)

Example 4-5

Calculate the potential energy of a 1500 pound rock precariously perched at the top of a ravine 120 feet above.

Solution:

$$PE = wh = 1500lb(120ft) = 180,000 \text{ lb-ft}$$

Example 4-6

Calculate the potential energy of a 900 kg rock precariously perched at the top of a ravine 40 meters above.

Solution:

$$PE = mgh = 900kg(9.81m/s^2)(40m) = 353,000J = 353kJ$$

When the raised object falls, its potential energy is converted into what is called **kinetic energy**, the energy of moving objects. The kinetic energy of a moving object depends on two things: the object's mass and the object's speed. Its speed is a major factor because the kinetic energy increases exponentially with the speed. Here we will use "KE" to represent kinetic energy.

In SI, when substituting units of mass in kg and units of speed squared in m^2/s^2, the units combine such that Joules result, units of work and energy. For the English system, mass in slugs must be multiplied to speed squared in ft^2/s^2, resulting in lb-ft, once again units of work and energy.

Kinetic Energy

$$KE = \frac{1}{2}mv^2$$

Example 4-7

Calculate the kinetic energy of a 5000 kg truck moving at a speed of 30 km/hr.

Solution:

$$v = 30km/hr = 8.33 \text{ m/s}$$

$$KE = \frac{1}{2}mv^2 = \frac{1}{2}(5000kg)(8.33ft/s)^2 = 173kJ$$

Example 4-8

Calculate the kinetic energy of a 11,000 pound truck moving at a speed of 30 mi/hr.

Solution:

$v = 30$ mi/hr $= 44$ ft/s

$m = w/g = 11{,}000 \text{lb}/32.2 \text{ ft/s}^2 = 342$ slugs

$KE = \dfrac{1}{2} mv^2 = \dfrac{1}{2} (342 \text{ slugs})(44 \text{mi/s})^2 = 331{,}000$ lb-ft

Note how for units to end in lb-ft, the weight in pounds must be converted to mass in slugs.

Conservation of Energy

Energy cannot be created or destroyed, but it can convert from one form of energy to another form. Sometimes the energy transformation is not very efficient, with "losses" unaccounted for, but the energy still exists somewhere in some form. More on efficiency in our next chapter.

Here we will study one situation where energy transforms from one form to another very efficiently. This is the situation where the potential energy of an elevated object is converted to the kinetic energy of the object when it falls.

An interesting point here is that all bodies regardless of their mass (disregarding air drag) fall at the same rate, the acceleration due to gravity. As a result mass cancels in the PE to KE transformation. Algebraically, we can divide both sides by "m", cancelling it from the equation.

> **Falling Object**
> **Conservation of Energy**
>
> **Potential Energy $\rightarrow$ Kinetic Energy**
>
> **PE $\rightarrow$ KE**
>
> $mgh \rightarrow \dfrac{1}{2} mv^2$

Example 4-9

Calculate the final velocity of an object that falls from a height of 10 meters.

Solution:

$PE = KE$

$\cancel{m}gh = \dfrac{1}{2} \cancel{m}v^2$

$gh = \dfrac{1}{2} v^2$

$v = \sqrt{2gh} = \sqrt{2(9.81 \tfrac{m}{s^2})(10m)} = 14.0$ m/s

James Prescott Joule (1818 – 1889)

James Prescott Joule was an English physicist and brewer, born in Salford, Lancashire. Joule studied the nature of heat, and discovered its relationship to mechanical work. This led to the law of conservation of energy, and to the development of the first law of thermodynamics. He worked with Lord Kelvin to develop the absolute scale of temperature. Joule also made observations of magnetostriction, and he found the relationship between the current through a resistor and the heat dissipated, which is now called Joule's first law.

The SI derived unit of energy, the joule, is named for James Joule.

Example 4-10

Calculate the final velocity of an object that descends the ramp shown.

Solution:

$$PE = KE$$

$$\cancel{m}gh = \frac{1}{2}\cancel{m}v^2$$

$$gh = \frac{1}{2}v^2$$

$$v = \sqrt{2gh} = \sqrt{2(9.81\tfrac{m}{s^2})(3m)} = 7.67 \text{ m/s}$$

The acceleration of the object down the ramp is less that the acceleration due to gravity, but it makes up the speed with having further to travel on the decline. As a result, only the height matters.

Example 4-11

Calculate the velocity of the pendulum when it swings down to its lowest point.

Solution:

$$v = \sqrt{2gh} = \sqrt{2(9.81\tfrac{m}{s^2})(10m)} = 14.0 \text{ m/s}$$

4.1 STUDENT EXERCISES   TRANSLATIONAL WORK

1. What is the unifying equation for work?

2. How does the unifying equation for work apply to translational mechanical energy systems?

3. How much work is needed to raise a 200 pound object to a height of 10 feet?

4. How much work does it take to raise a 1kN load to a height of 5 meters?

5. How much work is needed to raise 10 kg to a height of 20 meters?

6. By use of conservation of energy, calculate the maximum velocity an object will achieve once it has fallen sixty-five feet.

7. By us of conservation of energy, calculate the maximum velocity an object will achieve having fallen 22 meters.

8. What is the kinetic energy of a .223, 3.9 gram bullet traveling at 975 m/sec?

9. The mass of a baseball is about 145 grams. A good pitcher can throw the ball 100 mph (44 m/s) fastball. Calculate the kinetic energy of the ball.

10. The pitcher in problem # 9 throws the baseball straight up into the air at 100 mph. Using conservation of energy, calculate the maximum height the ball will achieve.

11. A 3 gram, 22 bullet has a velocity of 335m/s. Calculate its kinetic energy.

12. Using conservation of energy calculate the maximum velocity of an object that has fallen 25 feet.

Answers: 1. *the mover quantity of the energy system multiplied by the displacement quantity of the energy system* **2.** *the force multiplied by the distance* **3.** 2000 lb-ft **4.** 5000J **5.** 1960J **6.** 64.7ft/s **7.** 20.8m/s **8.** 1.85kJ **9.** 140J **10.** 98.7m **11.** 168J **12.** 40.1 ft/s

Our unifying principle states that the work in an energy system is the mover in the system multiplied by the displacement in the system. In rotational mechanical systems, the mover is the torque and the displacement is the angle of rotation.

Rotational Mechanical Work

$$W = \tau\theta$$

$$\tau = F\ell$$

Units of rotational work are once again lb-ft and Joules. But care must be taken to substitute radian measure in for theta (θ), the angular displacement. The reason for this is that torque already includes units of force and distance. Radians, a dimensionless ratio, will not introduce additional units into the equation. You may review radian measure in section 3.2 if needed.

Example 4-12

An electric motor provides 170 N-m of torque. How much work is done when the motor rotates through 500 revolutions?

Solution:

$$\theta = 500\,\text{rev}\left(\frac{2\pi\,\text{rad}}{1\,\text{rev}}\right) = 3140\text{ rad}$$

$$W = \tau\theta = (170\text{ N-m})(3140\text{ rad}) = 534\text{kJ}$$

Example 4-13

How much work is done by a power take-off shaft providing 300 ft-lb of torque through 1000 revolutions?

Solution:

$$\theta = 1000\,\text{rev}\left(\frac{2\pi\,\text{rad}}{1\,\text{rev}}\right) = 6280\text{ rad}$$

$$W = \tau\theta = (300\text{ ft-lb})(6280\text{ rad}) = 1.88 \times 10^6\text{ lb-ft}$$

---

1. How does the unifying equation for work apply to rotational mechanical systems?

2. How much work is needed to rotate a tiller shaft 80 revolutions at 34 ft-lb?

3. How much work is done on a flywheel when 120 N-m of torque rotates the flywheel 1500 revolutions?

4. Calculate the energy required to rotate a shaft 1000 revolutions with 6 ft-lb.

5. Explain why radian measure is necessary in $W = \tau\theta$.

Answer: 1. Torque times the angular displacement **2.** 17,100 lb-ft **3.** 1.13MJ **4.** 37,700 lb-ft **5.** Radian measure is a dimensionless ratio so does not introduce additional units, revolutions, in $\tau\theta$, in order to get consistent units of lb-ft or J

In fluid systems, work is done when a pressure difference causes liquids or gases to move. Following our unifying principle, the pressure (p) is multiplied to the displaced volume (V). Care must be taken for the units to be consistent thus producing the standard units of measurement for work, lb-ft in the English system and Joules in SI.

Winds blow because air moves from high pressure regions to low pressure regions. Remember that hydraulic systems uses liquids, usually oils, and pneumatic systems use a gas, often air.

Fluid Work
$W = pV$

Work Done by a Pump

A pump moves fluids. The work the pump does is the product of the pressure it produces and the volume of fluid it displaced. In the case of pumps, pressure is typically measured in psi (lb/in^2) in the English system, but this is usually converted to lb/ft^2 before substitution in the formula. Also, in the English system fluid volume is commonly measured in gallons, but this must be converted to ft^3. This is done so that feet cancel in the formula so that the resulting units become lb-ft.

Another concern is the extra step that must be taken to calculate the pressure given the height the pump must lift it. Review the stack formula ($p = \rho_w h$) and the water pressure depth rate 0.433 psi/ft in section 3.3 if needed.

Example 4-14

How much work does it take to pump 10,000 gallons of water to a height of 60 feet?

Solution:

$$p = (0.433 \text{psi/ft})(60 \text{ft}) = 25.98 \text{ psi} = (25.98 \text{psi})\left(\frac{144 \text{lb/ft}^2}{1 \text{psi}}\right) = 3741 \text{ lb/ft}^2$$

$$V = (10,000 \text{gal})\left(\frac{1 \text{ft}^3}{7.48 \text{gal}}\right) = 1337 \text{ ft}^3$$

$$W = pV = (3741 \text{ lb/ft}^2)(1337 \text{ ft}^3) = 5.00 \times 10^6 \text{ lb-ft}$$

Note how the units were set up so that ft^2 cancel completely in the pressure units and two of the ft^3 cancel in the volume units to leave lb-ft.

There is a shortcut solution for the problem of Example 4-14 that is worth studying. The problem can more easily be completed by treating the problem like a mechanical one, where the water is lifted up 60 feet like it were one weight. Yes, the water is pumped up a bit at a time, but in the end, a weight of water was lifted. The alternate method is called the **mechanical equivalent of fluid work**. This method is quicker because there are fewer conversions to make. The formula $W = Fd$ is used instead of $W = pV$. "F", the total weight of the fluid, is multiplied to the height "d".

Example 4-15

How much work does it take to pump 10,000 gallons of water to a height of 60 feet?

Solution:

F = weight of fluid = $(10,000 \text{ gal})(8.34 \text{ lb/gal}) = 83,400 \text{ lb}$

$W = Fd = (83,400 \text{ lb})(60 \text{ ft}) = 5.00 \times 10^6 \text{ lb-ft}$

Example 4-16

How much work does it take to pump 10,000 gallons of **seawater** to a height of 60 feet?

Solution:

SG of seawater = 1.04

F = weight of fluid = $(10,000 \text{ gal})(8.34 \text{ lb/gal})(1.04) = 86,740 \text{ lb}$

$W = Fd = (86,740 \text{ lb})(60 \text{ ft}) = 5.20 \times 10^6 \text{ lb-ft}$

Note how the specific gravity of seawater is multiplied into the equation. Seawater is 1.04 times heavier than water, so requires 1.04 times more work/energy to lift.

Example 4-17

How much work does it take to pump 20,000 L of water to a height of 15 m?

Solution:

1L water = 1kg = 9.81N

F = weight of fluid = $(20,000 \text{ L})(9.81 \text{N/L}) = 196 \text{ kN}$

$W = Fd = (196 \text{kN})(15 \text{ m}) = 2940 \text{ kJ}$

Work Done On/By a Piston

In our chapter on the mover quantities we learned that the pressure in a cylinder (piston) is the force on the piston shaft (F) divided by the cross-sectional area (A) of the cylinder, $p = F/A$. To calculate the fluid work of the cylinder, $W = pV$, the change in volume of fluid inside the cylinder must be determined. The piston head moves inside the piston, so the change in volume is the shape of a right circular cylinder. The formula for the volume of a right circular cylinder is $V = \pi r^2 h$. We studied the volume of a right circular cylinder in section 1.3, so step back and review this if needed. Also, the letter delta (Δ) is commonly used in these applications to indicate a "change in" volume (ΔV), this to distinguish it from the total volume of the cylinder (V).

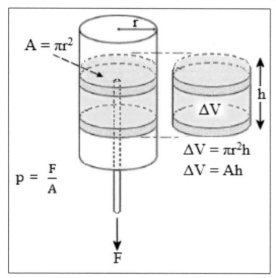

Note in the diagram how the cylinder head slides from one point in the cylinder to another. In a hydraulic application, this head movement displaces an incompressible liquid (such as oil) into or out of the cylinder. The amount of displaced liquid, depicted in yellow in the diagram, takes the shape of a right circular cylinder, calculated using $\Delta V = \pi r^2 h$ with "h" representing how far the cylinder head moved.

In the case of a pneumatic system, the air (or other gas) need not exit or enter the cylinder because the gas can be compressed. As a result, the mass or amount of air doesn't change, but the volume of the air changes, being compressed or expanded by the cylinder. The change in volume is calculated the same way as in the hydraulic case using $\Delta V = \pi r^2 h$.

You probably noticed the "on/by" when referring to work as applied to pistons. Pistons can be used in two basic ways. The first way is that fluid can be pumped into the cylinder, moving the piston head, and producing a force at the shaft. Fluid pressure in, mechanical force out. This converts fluid work to mechanical work, like the pistons on tractor implements such as loaders, a

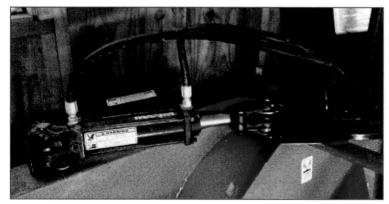

snow thrower chute control, etc. The second way is the other way around, where mechanical force is applied at the shaft, compressing or displacing fluid in the cylinder. Mechanical force in, fluid pressure out. We will investigate both of these modes of operation in laboratory activities.

Example 4-18

A 6 inch diameter hydraulic cylinder moves a piston 10 inches while moving a mechanical load. The fluid pressure is 800 psi.
Calculate the work done by the cylinder.
Solution:

$r = 3$ inches

$\Delta V = \pi r^2 h = 3.14(3\text{in})^2(10 \text{ in}) = 283 \text{ in}^3$

$W = pV = (800\text{lb/in}^2)(283 \text{ in}^3) = 226,400 \text{ lb-in}$ or $18,900 \text{ lb-ft}$

Units of lb-in are legitimate units of work and acceptable as an answer here, but lb-in can easily be converted to lb-ft simply by dividing by 12.

Example 4-19

A 10 centimeter diameter hydraulic cylinder moves a piston 24 centimeters while moving a mechanical load. The fluid pressure is 75 kPa.
Calculate the work done by the cylinder.
Solution:

$r = 5 \text{ cm} = 0.05 \text{ m}$

$h = 24 \text{ cm} = 0.24 \text{ m}$

$p = 75\text{kPa} = 75,000\text{Pa} = 75,000 \text{ N/m}^2$

$\Delta V = \pi r^2 h = 3.14(0.05\text{m})^2(0.24\text{m}) = 1.88\text{x}10^{-3} \text{ m}^3$

$W = pV = (75,000\text{N/m}^2)(1.88\text{x}10^{-3} \text{ m}^3) = 141 \text{ J}$

1. How does the unifying equation for work apply to fluid systems?

1. How much work in needed to pump 1000 gallons up from a well 140 feet deep?

2. What is the maximum fluid displacement of a four inch diameter cylinder with a 12 in stroke? ("Stroke" is the maximum length of the piston shaft, the maximum distance the cylinder can move an object.)

3. How much energy is required to pump 5,000 liters to a height of 40 meters?

4. A piston with a diameter of three inches operating at 400 psi displaces 85 cubic inches of oil. Calculate the work done.

5. A small 2 centimeter diameter cylinder operating at 5kPa moves a mechanical load 8.5 centimeters. Calculate the work done.

> **Answers: 1.** *the pressure multiplied by the volume of fluid displaced* **2.** 1.17×10^6 lb-ft
> **3.** 151 in^3 **4.** 1.96MJ **5.** 34,000 lb-in or 2830 lb-ft **6.** 0.133J

Following our unifying principle for work, electrical work occurs when the electrical mover, the voltage (E), displaces charge (q).

Earlier we learned that voltage is measured in volts (V) and charge is measured in coulombs (c). We also learned that current (I) is the rate at which charge flows, or $I = q/t$, where one ampere (A) equals one coulomb per second (c/s).

There are no English units of measurement for electrical work. Volts, amps, and coulombs are generally considered to be SI. So there is one system of units for electrical quantities.

When it comes to electrical work and energy, scientist Coulomb experimentally determined what is called the mechanical equivalent of electrical work. He figured out how many electrons, the charge, that must be displaced by one volt to be equal of one Joule of energy. So the size of the unit coulomb was established in order for electrical systems to be consistent with the other energy systems.   As a result, the formula $W = Eq$ produces standard metric units of work and energy, the Joule.

> **one volt-coulomb = 1 one Joule**

Example 4-20

How much work/energy can a 12V battery when fully charged with 7500 coulombs of charge?

Solution:

$$W = Eq = 12V(7500c) = 90,000J = 90kJ$$

But we have a problem if we're to do electrical work laboratory activities; there is no such thing as a coulomb meter. But we do have ammeters. Since $I = q/t$, then $q = It$, so if we can measure the current (I) with our meter and the time interval (t) with a timer, these can be multiplied to get the charge in coulombs. By substituting It in for q in $W = Eq$, we have a formula for electrical work that contains three things we can measure, voltage, current, and time.

> **W = EIt**
>
> **1 Joule = 1 volt-amp-sec**

Example 4-21

How much work/energy is produced by a 12V electrical generator when 25 amperes flows for one hour?

Solution:

$t = 1$ hr $= 3600$ sec
$W = EIt = 12V(25A)(3600s) = 1.08 \times 10^6 \, J = 1.08$ MJ

*To get Joules, time must be in seconds because an ampere is one coulomb per **second**.*

Example 4-22

How much work/energy is consumed by a 12V electrical motor that draws 10 amperes from a power source for one minute?

Solution:

$t = 1$ min $= 60$ sec
$W = EIt = 12V(10A)(60s) = 7200 \, J$

4.4 STUDENT EXERCISES — ELECTRICAL WORK

1. How does the unifying equation for work apply to electrical systems?

2. How much work can be performed by a 12 volt battery containing 10,000 c?

3. A 12 volt battery provides 65A to the engine's started for 2 seconds. Calculate the work done by the battery.

4. How much energy is consumed by a 110V load drawing 16A for 24 hours?

5. The displacement quantity in electrical systems is the _____.

6. The unit of measurement for electrical displacement is _____.

7. The rate at which charge is displaced is the _____.

8. The unit of measurement for the rate at which charge is displaced is _____.

9. Calculate the electrical energy consumed by a 220V electric lift that draws 10A for fifteen seconds.

Answers: 1. *the voltage multiplied by the charge* **2.** 120kJ **3.** 1560J **4.** 1.52×10^8J
5. *charge* **6.** the coulomb **7.** current **8.** the ampere **9.** 33kJ

Our unifying principle does well to unify four energy systems. This one powerful idea regarding work can be applied to translational mechanical, rotational mechanical, fluid, and electrical systems. But we have an exception. The unifying principle for work does not apply to thermal systems. In thermal systems, the work/energy is the heat displaced itself. As a result, Btu's and calories can be directly converted to lb-ft and Joules. We learned how to calculate heat energy in Section 1.5, our study of thermal displacement. In brief review, two equations, one for latent heat and one for sensible heat, are used to determine thermal work:

Sensible Heat
$Q = mc\Delta T$

Water
Latent Heat of Fusion $Q = mH_f$
Latent Heat of Vaporization $Q = mH_v$

1 Btu = 252.0 cal = 77.79 lb-ft = 1055 J

Water
$H_f = 144$ **Btu/lb = 79.8 cal/g**
$H_v = 970$ **Btu/lb = 540 cal/g**

Example 4-23

  a. Convert 15,000 Btu to lb-ft

    Solution:

$$(15{,}000 \text{ Btu})\left(\frac{77.79 \text{ lb}-\text{ft}}{1 \text{ Btu}}\right) = 1.17 \times 10^6 \text{ lb-ft}$$

  b. Convert 175,000 calories to Joules

    Solution:

$$(175{,}000 \text{ cal})\left(\frac{1055 \text{ J}}{252 \text{ cal}}\right) = 7.33 \times 10^5 \text{ J}$$

  c. Convert 1.22×10^8 J to Btu.

    Solution:

$$(1.22 \times 10^8 \text{ J})\left(\frac{1 \text{ Btu}}{1055 \text{ J}}\right) = 1.16 \times 10^5 \text{ Btu}$$

Example 4-24

How much work/energy is required of a water heater that brings 45 gallons of water from 68°F to 170°F?

Solution:

$$\Delta T = 170°F - 68°F = 102 \ F°$$

$$W = Q = mc\Delta T = (45 \ \cancel{gal})(8.34 \ \cancel{lb/gal})(1 \ Btu/\cancel{lb\text{-}F°})(102 \ \cancel{F°}) = 38{,}300 \ Btu$$

$$W = (38{,}300 \ \cancel{Btu})(77.79 \ lb\text{-}ft/\cancel{Btu}) = 2.98 \times 10^6 \ lb\text{-}ft$$

Although leaving the answer in Btu units would be acceptable here, we are asked to calculate the work/energy. Technically, work/energy is measured in lb-ft or Joules.

4.5 STUDENT EXERCISES   THERMAL WORK

1. How the unifying equation for work apply to thermal systems?

2. How much energy is needed to melt one kilogram of ice that is zero degrees centigrade?

3. How much energy is needed to completely boil away one gallon of water that is initially seventy degrees Fahrenheit?

4. How much thermal work is required to heat 20 gallons of water from 65°F to 180°F?

5. A 500 gram chunk of ice is −20°C. How much heat energy is needed to melt the ice, then completely boil it away to vapor?

Answers: 1. *It doesn't. Thermal work is equal to heat energy displaced.* **2.** 79.8 kcal
3. 5690 Btu **4.** 19,200 Btu **5.** 365kcal

All machines typically transform one form of work/energy into another. For example, an electric water pump converts electrical work at its input into fluid work at its output. In real-life machines, the output work is always less than the input work because some of the input work must be used to overcome undesired but unavoidable losses. In mechanical systems these losses appear as friction between all the moving parts. In fluid systems, pipes have resistance similar to the electrical resistance we studied earlier. As a result, all machines loose some of the energy pit into them because of these energy losses.

Machines are often rated by their efficiency. **Efficiency** is a dimensionless ratio of output work to input work and usually expressed as a percent. Comparing the useful work accomplished by a machine to the amount of work that must be put into it quantifies how well a machine operates with respect to losses. Efficiency is sometimes abbreviated in formula with "eff", but in technical literature efficiency is commonly abbreviated with Greek letter eta, η. (Do not confuse eta with mu, μ.)

> **Efficiency**
>
> $$\eta = \frac{W_{out}}{W_{in}}$$

Example 4-25

What is the efficiency of a mechanical pulley system that produces 1200 lb-ft of work lifting an engine when 1500 lb-ft of work must be put in?

Solution:

$$\eta = \frac{W_{out}}{W_{in}} = \frac{1200 \, \cancel{lb-ft}}{1500 \, \cancel{lb-ft}} = 0.800 = 80.0\%$$

Example 4-26

A 22% efficient water pump must accomplish 140MJ of work. How much electrical energy must be put into the system?

Solution:

$$W_{in} = \frac{W_{out}}{\eta} = \frac{140 \, MJ}{0.22} = 636 \, MJ$$

Note how the formula is manipulated to calculate W_{in}, where the W_{out} must be divided by the efficiency. Remember W_{in} must always be greater than W_{out}.

LABORATORY ACTIVITY PULLEYS **L13**

Student Name(s) _____

_____

_____

_____

Work Done by Pulleys

MAIN IDEAS

- Pulleys (block and tackle) often make work easier by amplifying force.
- Pulleys do not increase or amplify work because there is a trade-off; they increase the output force at the expense of the output distance.
- Mechanical advantage is usually expressed as the ratio of the "Force Out" to "Force In" or

$$MA = \frac{F_{out}}{F_{in}}$$

- Efficiency is a comparison of "Work Out" to "Work In" or

$$eff = \frac{W_{out}}{W_{in}}$$

- Efficiency is usually expressed as a percent.

Combinations of pulleys, called "block and tackle", are used to make the job of lifting heavy objects easier. Even when different types of motors move heavy loads, pulley systems are often still used.

Technically speaking, work is not done unless a mover quantity causes some form of displacement. In linear mechanical systems, this means a force must move an object a distance. Both force *and* distance must be present at the same time for work to be done.

In pulley systems, linear mechanical work appears at both the input and output of the system where the input and output work can be expressed as $W_{in} = F_{in}d_{in}$ and $W_{out} = F_{out}d_{out}$. The pulley system does not produce work, it only changes the combination of force and distance. In most cases a small force is put into the system while a large force appears at the output, but for this to occur, a price must be paid: distance is sacrificed. As a result, a small force moving a large distance can be equal to a large force moving a small distance, with their products, work, being equal (at least under ideal conditions).

Pulley systems have a mechanical advantage (MA), that is, a number representing how much more force appears at the output than what is put in. This is found by simply dividing the output force by the input force.

All systems operate at efficiencies less than 100% because of losses such as from friction. Efficiency is a measure of these losses, the ratio of output work to input work, usually expressed as a percent. For example, if only half of the work put into the system appears as useful work at the output of the system, the efficiency would be 50%.

Mechanical Advantage
$MA = \dfrac{F_{out}}{F_{in}}$

Although more *force* can appear at the output, there can never be more *work* at the output because all machines and systems loose energy because of inevitable losses such as from friction. In fact, as the mechanical advantage increases, the efficiency decreases because of the addition of more friction from more moving parts.

Most systems have what is called an **ideal mechanical advantage** (IMA), which is the ratio of output to input force that the system would have if there were no energy losses, or under ideal conditions. But no system is without energy losses, so the system has what is called the **actual mechanical advantage** (AMA), the ratio of output to input force that the system actually gets under real conditions and including losses.

Pulley IMA by Counting Supporting Ropes

There's a neat trick that can be used to determine the ideal mechanical advantage of any pulley system by inspection; simply count the supporting ropes.

Remember our discussion of how vectors add. Here we have one vector pointing downward, the weight of the load, and three upward, the three supporting ropes. The three ropes each support one third of the load. As a result, there is one third the load weight of tension in the rope, and in the example at right, an ideal mechanical advantage of three.

This method is called **sectioning**. It is done by visually cutting off the moveable pulley system from the fixed pulley system. The IMA is *not* determined by counting pulleys, but by counting supporting ropes on the moveable part of the system.

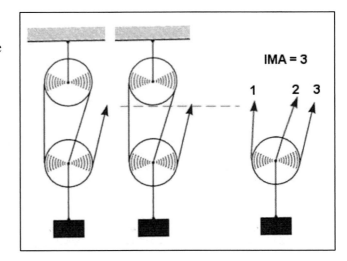

258

Care must be taken as to whether the end of the rope, that is, the part being pulled on, is counted or not. It only counts when the hand is pulling up, a supporting rope. If the pulley system is set up such that the hand pulls downward, this is not a *supporting* rope and is not counted in the IMA count. Note how in the example at right the rope being pulled downward is not a factor in the IMA count.

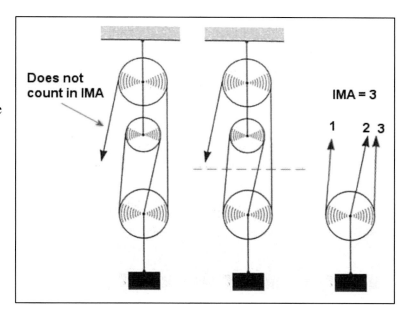

Equipment

Spring scales

Pulleys

String

Two meter sticks

Set of small-capacity slotted weights

50 gram weight hanger

Hook collars

Support stand, rods, and clamps

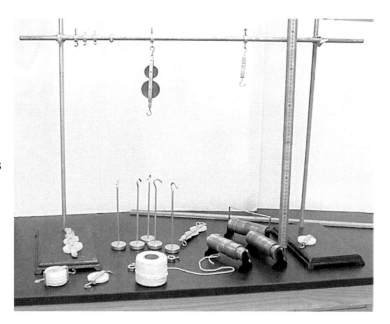

Procedures

Initial Setup

Use two support stands and two clamps to suspend a support rod horizontally at a convenient height as shown in the figure at right. The hook collars slip over the rod and provide places from which pulleys are suspended.

Part 1: MA = 1

1. Set up a single, fixed pulley as shown. Attach a weight (F_{out}) to the load end of the cord. Record F_{out} in the Data Table.
2. Attach the spring scale to the other end of the cord so that you **pull on the hook of the scale. Zero the spring scales,** then pull on the hook and measure the force (F_{in}) it takes to start the weight moving slowly upward at constant speed. Record F_{in} in the Data Table.
3. Pull on the scale hook and cause the weight to move up slowly some distance d_{out}. Use a meter stick to measure the output distance d_{out}. Record d_{out} in the Data Table.
4. At the same time that the weight is moving up a distance d_{out}, measure the input distance d_{in} that the hand moves at the spring scale. A certain amount of team work will make it possible to record the distances simultaneously.

Part 2: MA = 2

Arrange the pulley system shown in the figure at right. Use the same weight as in Part 1. Measure the same quantities of force and distance. Record these values in the Data Table. Note that although a smaller force is required to move the same weight, more distance is needed at the input.

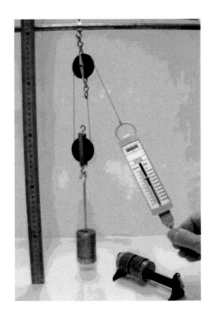

Part 3: MA = 3

Arrange the pulley system shown in the figure at right. Use the same weight as in Part 1. Again measure the same quantities of force and distance. Record these values in the Data Table. Note that although yet a smaller force is required to move the same weight, even more distance is needed at the input.

Part 4: MA ≥ 4

Arrange any pulley system of your own design having MA ≥ 4. Use the same weight as in Part 1. Again measure the same quantities of force and distance.. Record these values in the Data Table. Note that although yet a smaller force is required to move the same weight, even more distance is needed at the input.

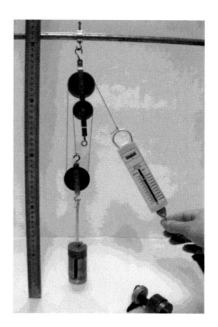

Data Table

MA	F_{out}	d_{out}	F_{in}	d_{in}	W_{out}	W_{in}	η
1							
2							
3							
≥4							

Questions

1. What is the unifying equation for work?

2. How does the unifying equation for work apply to the above linear mechanical pulley systems?

3. The pulley system does not produce more work at the output, but more force. What must compensate for this?

4. Explain the difference between the mechanical advantage of the pulley system and the efficiency of the pulley system.

5. As the mechanical advantage is increased in a mechanical system, usually the efficiency can be expected to reduce. Why is this so? Do your calculations support this expectation?

6. It is often mistakenly said that a pulley system produces more energy at its output than what is put in. Explain how this common misconception is wrong.

Student Name(s)_____

_____

_____

Work Done by a Wheel and Axle

Discussion

We use simple machines every day to help make difficult tasks easier. Simple machines provide a trade-off between the force applied and the distance over which the force is applied. A wheel and axle is a simple machine made of two parts, an axle (or rod) attached to the center of a larger wheel. The wheel and axle move together - effort applied to the wheel turns the axle, or effort applied to the axle turns the wheel. Wheel and axles are used to lift or move loads. A wheel and axle can produce a gain in either effort or distance, depending on how it is used. When the force is applied to the wheel in order to turn the axle, force is increased and distance and speed are decreased. When the force is applied to the axle in order to turn the wheel, force is decreased and distance and speed are increased. Here in this exercise, consider the outer wheel as the input side and the inner axle as the output side.

The mechanical advantage of the system can be calculated by way of either of three ratios, each upon experimentation approximately equal:

$$MA = \frac{\text{input distance}}{\text{output distance}}$$

$$MA = \frac{\text{radius of the outer wheel}}{\text{radius of the inner axle}}$$

$$MA = \frac{\text{output force}}{\text{input force}}$$

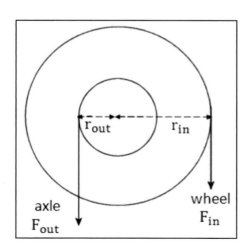

The wheel and axle is, in a way, two levers working together where the wheel and axle radii are the lever arms. The system is balanced when there is equal torque to either side. When balanced and not rotating, the net torque is zero. As a result, like a lever, a small force on one side can move a bigger weight by sacrificing distance.

Torque = force times the lever arm or $\tau = Fl$

The efficiency (η), as usual, is the ratio of the output work to the input work, usually expressed as a percent:

$$\eta = \frac{W_{out}}{W_{in}}$$

Apparatus:

Support stand and clamp

Two weight hangers

Set of slotted weights

Wheel and axle apparatus

String and scissors

Ruler

Set-up

1. Mount the wheel and axle apparatus to the support stand with the clamp as shown.
2. Tie a string to the inner disc (axle), preferably the inner most, using the small hole provided. This is done by making a large knot behind the hole.
3. Tie a second string to one of the outer wheels (the wheel) preferably the furthest out.
4. Tie weight hangers to the strings as shown.
5. Apply weights until the system is balanced and does not turn. For equilibrium, there will necessarily be more weight to one side.

Measurements

1. Using the ruler, measure the radii of the larger wheel and also the smaller axle. Record these in the provided data table.
2. Record the weight values, one for the wheel weight (outer) and one for the axle weight (inner) in the table provided.
3. With the axle (inner) weight at a high position, pull it down a convenient distance. Measure and record this axle distance and also the resulting wheel weight distance in the table provided.

Data Table

	Radius	Weight	Distance Moved
Wheel			
Axle			

Calculations

1. Mechanical Advantage $= \dfrac{\text{radius of the outer wheel}}{\text{radius of the inner axle}} = \dfrac{(\quad\quad)}{(\quad\quad)}$

   MA = ─────────────

2. Mechanical Advantage $= \dfrac{\text{input distance}}{\text{output distance}} = \dfrac{(\quad\quad)}{(\quad\quad)}$

   MA = ───────────── =

3. Mechanical Advantage $= \dfrac{\text{output force}}{\text{input force}} = \dfrac{(\quad\quad)}{(\quad\quad)}$

   MA = ───────────── =

4. Wheel torque = wheel weight x wheel radius = (   )(   ) = _____

5. Axle torque = axle weight x axle radius = (   )(   ) = _____

6. Net torque = ± wheel torque ± axle torque = ±(   ) ± (   ) = _____

7. Efficiency $= \square = \dfrac{W_{out}}{W_{in}} =$ ───────────── = _____

Wrap-Up

1. Theoretically, when balanced, the net torque should be zero. Do your calculations support this?

2. The three ways by which mechanical advantage is calculated should agree. Do your calculations support this?

3. Explain what is meant by the mechanical advantage of a simple machine.

4. Explain why it is possible to get more force out of a mechanical system but impossible to get more work.

5. Simple machines tend to operate efficiently because there are few moving parts. Explain why efficiency calculations should always result in less than 100%.

6. Explain how the wheel and axle is a form of lever.

Student Name(s)_____

_____

_____

Work Done by a Winch

Main Ideas

- A winch is a mechanical machine that makes work easier.
- Work done on a winch is work done by an applied torque. This torque causes rotation.
- A winch converts rotational mechanical work input into linear mechanical work output.
- The efficiency of a winch can be calculated from the ratio Work Out / Work In, where Work Out is the lifted weight and Work In is the rotational work put in.
- The mechanical advantage of a winch is the ratio of the output weight lifted to the input force that creates the torque.
- A winch is a mechanical device made of a cylindrical drum and a handle that rotates. Winches use a rope or cable to haul in a load.
- Efficiency of a winch is always less than 100%.

Equipment

Winch assembly, including rope and string

Heavy-duty support stand

Spring scale

Large-capacity weights

Meter stick or tape measure

Spring scale, 0-10 Newton

Discussion

In terms of work, the winch converts rotational mechanical work into linear mechanical work. Not all the work applied appears as output work; there are losses. According to the unifying principle for work, work is calculated by multiplying the force-like quantity to the displacement quantity. The input and output work of the winch is calculated using this principle.

The *Input Work* is mechanical rotational. In rotational systems, the force-like quantity is the *torque*, symbolized with the Greek letter tao (τ). The rotational displacement quantity is the *angle* the winch disc turns, symbolized by the Greek letter theta (θ). According to the unifying principle, rotational work is then expressed by the formula

$$W = \tau\theta.$$

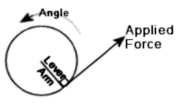

In a winch, your hand *applies the force* to crank the handle. This is measured with the spring scale. The *lever arm* is the length of the crank from the handle to the axle or the radius of the disc. The *torque applied* is the force times the lever arm or $\tau = Fl$. To find the Work In, multiply this *torque* by the *angle turned* in radians.

As with all mechanical machines, the *actual mechanical advantage* of the winch is the ratio of the output weight lifted to the input force creating the torque or

$$MA = \frac{Output\,Force}{Input\,Force}.$$

In this lab the Output Force is the weight moved (w) and the input force is the force applied to the disc (*F*).

The *Output Work* is linear mechanical. Applying the unifying principle, the work done is W=Fd. Here the force-like quantity is defined as the load *weight* (F) and the displacement is the *distance* (d) or

$$W = Fd$$

The *efficiency* of the system is the ratio of the output work to the input work expressed as a percent.

$$\text{Efficiency} = \text{Work Out/ Work In}$$

Procedures

Part 1: Set-up

1. Attach the winch assembly to the heavy-duty support stand.
2. Unreel enough cord to connect to the cord to 10 kg weight on the floor.
3. With string wrapped around the winch disc, attach the spring scale to the free end so that you can pull to turn the disc, measuring the applied force.
4. Read the remaining steps before continuing.

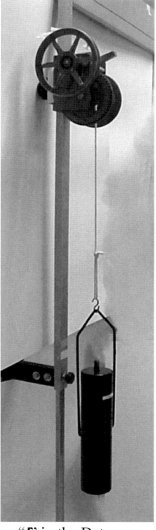

Part 2: Collecting Data

1. Measure the radius of the winch disc in meters. Record this value as "r" in the Data Table.
2. Record the total output load (in Newtons) as "F_o" in the Data Table.
3. Keeping the applied force on the disc perpendicular to the lever arm, pull the spring scale with just enough force to move the load a short distance at constant speed. Record this value as "F_i" (in Newtons) in the Data Table.
4. Return the load to its starting point. Keep the cord taut between the load and winch in order to minimize cord stretching. Raise the load half a meter or more by turning the disc, counting the number of revolutions. Record the number of revolutions as "θ" in the Data Table.
5. Record the height the load was lifted as "d" (in meters) in the Data Table.

Part 3: Calculations

Complete the Calculations Table by applying the appropriate formulae.

Data Table

Weight of Load F_o (Newtons)	Lever Arm l (meters)	Pulling Force F_i (Newtons)	Angle Turned θ (revs)	Load Distance d (meters)

Calculations Table

Torque	Angle	Work In	Work Out	% Efficiency
$\tau = Fl$ (in N-m)	$\theta = 6.28(\text{revs})$ (in radians)	$W_i = \tau\theta$ (in Joules)	$W_o = F_o d$ (in Joules)	$\eta = W_o/W_i$

Wrap-Up

1. Why is the output work less than the input work?

2. If the winch is less than 100% efficient, why do we still use them to move weights?

Student Challenge

1. In terms of force applied, calculate the *mechanical advantage* of the winch.

2. Using the value for efficiency calculated for this winch, determine the amount of input work required to lift a 1000 pound load a distance of 2 feet.

3. For the situation in Student Challenge #2 above, what input force would have to be applied to move the given load as described? (Hint: Calculate actual mechanical advantage using data from the data table.)

4. Explain radian measure and why its use is necessary when calculating rotational work.

WORK DONE ON A PISTON L16

Student Name(s)_____

_____

_____

Work Done ON a Piston

Lab Objectives

When you've finished this lab, you should be able to do the following:

- Calculate the mechanical work done on a piston by a force that moves the piston.
- Calculate the fluid work done by a piston on a gas that's compressed by the piston.
- Measure the mechanical and fluid work done by a piston that compresses gas in a cylinder.

Main Ideas

- In a closed cylinder, a fluid under pressure can do work on a piston.
- In a closed cylinder, a piston may do work on a fluid as it moves it.
- Work is done on a gas to squeeze it into a smaller volume.
- When a piston moves a liquid in a closed system, the volume of the liquid does not change. A liquid is virtually incompressible.
- The volume of a cylinder is the cross-sectional area of the cylinder times the length of the cylinder

In **hydraulic cylinders**, a liquid under pressure moves a piston. In turn, the piston moves a mechanical load and does useful work. In steam engines, the steam (which is under high pressure) moves a piston back and forth and does useful work. In gasoline engines, a piston does work on a mixture of air and gasoline vapor while it compresses the gas into a smaller volume.

In this lab, we'll study the work done on a **pneumatic cylinder** while a weight (force) moves the piston head such that it compresses the gas in the cylinder. We will use a simple piston and cylinder arrangement that "squeezes" gas into a smaller volume. We'll find the **Work done by** using the equations $W = Fd$, the mechanical work applied, and $W = pV$, the fluid **work done on** the gas.

When the piston moves inside the cylinder, the piston compresses the gas into a smaller volume. This decrease in volume is equal to the volume of air displaced by the piston. This volume is calculated by multiplying the cross-sectional area of the cylinder to the change in the height of the piston. The cross-sectional area is calculated using $A = \pi r^2$. The volume is then $\Delta V = Ah$ where A is the area and h is the height the weight moved (which is the same as the height of the compressed volume in the cylinder).

Equipment

Small air cylinder/piston, $1\frac{1}{16}$ inch diameter (unless otherwise indicated by your instructor)

Large-capacity set of slotted weights

Meter stick or ruler

Clamps

Pressure gage

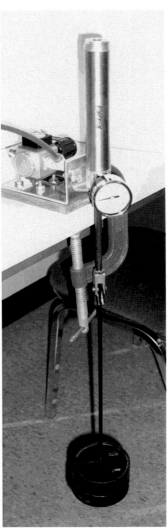

Procedures

Part 1: Collecting Data

1. Record the value of the diameter of the air piston in the data table. Calculate the cross-sectional area of the cylinder and record this value in the data table.
2. Clamp the support board and cylinder securely to the lab bench.
3. Set the piston at its highest for the initial starting position.
4. Connect a pressure gage to the cylinder. (The connector has a tab which must be depressed to connect and disconnect the gage.)
5. Hang a weight on the shaft that approximately produces a quarter-scale reading on the gage.
6. Record this weight in the Data Table under Trial #1.
7. Record the gage reading in the Data Table under Trial #1.
8. Measure the distance the piston and weight moved. Record this distance in the Data Table under Trial #1.
9. Repeat steps 5 through 9, but hang additional weight so that it approximately produces a three-quarter scale gage reading, recording the values in the Data Table under Trial #2.
10. Remove the weight from the piston.

Part 2: Calculations

1. Calculate the work done on the cylinder using $W = Fd$ where F is the weight in pounds and d is the distance the weight moved in inches. Record this value in the Mechanical Work Calculations Table

2. Calculate the pressure in the cylinder using $p = F/A$ where F is the weight in pounds and A is the area in square inches. Record this value in the Fluid Work Calculations Table.

3. Calculate the change in volume using $\Delta V = Ah$ where V is the volume, A is the area and h is the change in height. The change in height is equal to the distance (d). Record this value in the Calculations Data Table.

4. Calculate the Fluid Work using $W = pV$ where p is the pressure and V is the volume. Record this value in the Calculations Data Table.

Data Table

Piston Diameter = _____ in   Piston Area = _____ in^2

Trial #	Weight on Piston (lbs)	Distance Weight Moved (in)	Gage Reading (psi)
1			
2			

Calculations Tables

Mechanical Work

Trial #	Force on Piston F (lb)	Movement of Piston d (in)	Mechanical Work W =Fd (lb-in)	Mechanical Work (lb-ft)
1				
2				

Fluid Work

Trial #	Calculated Pressure p = F/A (psi)	Change in Volume $\Delta V = Ah$ (in³)	Fluid Work W = pV (lb-in)	Fluid Work (lb-ft)
1				
2				

Wrap-Up

1. What is the unifying equation for *work*?

2. How does the unifying equation for work apply to fluid systems?

3. In this lab you computed the pressure in the cylinder using p = F/A. You also recorded the pressure using a mechanical pressure gage. How do these two values compare? Are they the same? If they are different in value, which of the two is most likely to be correct?

4. Compare the results for **mechanical work** to **fluid work** for each of the two trials. Are they equal? Should they be?

WORK DONE BY A PISTON L17

Student Name(s)_____

_____

_____

Work Done By a Piston

Lab Objectives

When you've finished this lab, you should be able to do the following:
- Calculate the mechanical output work done by piston lifting a weight.
- Calculate the fluid input work done on a piston when lifting a weight.
- Calculate the efficiency of the pneumatic cylinder.

Main Ideas

- In a closed cylinder, a pneumatic cylinder can convert fluid work to mechanical work.
- Mechanical work is done by the cylinder when pressurized air displaces the cylinder head.
- Air is displaced into the cylinder.
- The volume of a cylinder is the cross-sectional area of the cylinder times the length of the cylinder.
- The efficiency of the cylinder is the ration of the output mechanical work to the input fluid work.

In **hydraulic cylinders**, a liquid under pressure moves a piston. In turn, the piston moves a mechanical load and does useful work. In this lab, we'll apply pressurized air to the pneumatic cylinder, moving the cylinder head and thus creating mechanical force. We will compare the fluid work in to the mechanical work out. We'll calculate the **Work** by using the equations $W_o = Fd$ and $W_i = pV$.

When the piston moves inside the cylinder, air forced into the cylinder increases the volume of air in the cylinder. This increase in volume is in the shape of a right circular cylinder. This volume is calculated by multiplying the cross-sectional area of the cylinder to the change in the height of the piston. The cross-sectional area is calculated using $A = \pi r^2$. The volume is then $\Delta V = Ah$ where A is the area and h is the height the weight moved (which is the same as the distance the cylinder head moved).

$$\eta = \frac{W_{out}}{W_{in}} = \frac{Fd}{pV}$$

Equipment

Small air cylinder/piston, $1\frac{1}{16}$ inch diameter
Large-capacity set of slotted weights
Meter stick or ruler
Clamps
Pressure gage
Compressed air supply (compressor or hand pump)

Procedures

Part 1: Collecting Data

1. Record the value of the diameter of the air piston in the data table. Calculate the cross-sectional area of the cylinder and record this value in the data table.
2. Clamp the support board and cylinder securely to the lab bench.
3. Set the piston at its lowest for the initial starting position.
4. Disconnect the pressure gage from the cylinder and put it aside. (The connector has a tab which must be depressed to connect and disconnect the gage.)
5. Hang weight on the shaft, up to as much as 15 kg (don't hold back on the weight).
6. Record this weight in the Data Table under Trial #1.
7. Connect the air supply tube from the lab station to the pressure regulator. (See instructor for specifics.)
8. Turn the regulator fully counterclockwise to the full off position.
9. Carefully open the air supply valve.
10. Very slowly and gently turn the regulator clockwise to allow air pressure to the cylinder applying barely enough pressure to lift the weight.
11. You will reach a point where there is enough air pressure to lift the weight. (The weight will probably move upward full stroke.)
12. Record the gage reading in the Data Table under Trial #1.
13. Measure the distance the piston and weight moved. Record this distance in the Data Table under Trial #1.

Part 2: Calculations

1. Calculate the work done on the cylinder using W = Fd where F is the weight in pounds and d is the distance the weight moved in inches. Record this value in the Mechanical Work Calculations Table

2. Calculate the pressure in the cylinder using P = F/A where F is the weight in pounds and A is the area in square inches. Record this value in the Fluid Work Calculations Table.

3. Calculate the change in volume using ΔV = Ah where V is the volume, A is the area and h is the change in height. The change in height is equal to the distance (d). Record this value in the Calculations Data Table.

4. Calculate the Fluid Work using W = PV where P is the pressure and V is the volume. Record this value in the Calculations Data Table.

Data Table

Piston Diameter = _____ in   Piston Area = _____ in^2

Weight (lbs)	Distance Weight Moved (in)	Gage Reading (psi)

Calculations Table

Mechanical Work

Force (weight) F (lb)	Movement of Piston d (in)	Mechanical Work W = Fd (lb-in)	Mechanical Work (lb-ft)

Fluid Work

Pressure (psi) Calculated	Change in Volume ΔV = Ah (in^3)	Fluid Work W = pV (lb-in)	Fluid Work (lb-ft)

Efficiency $\quad \eta = \dfrac{W_{out}}{W_{in}} = \dfrac{Fd}{pV} \qquad \eta = \underline{\hspace{3cm}}$

Wrap-Up

1. What is the unifying equation for *work*?

2. How does the unifying equation for work apply to fluid systems?

3. In this lab you measured the applied pressure using the regulator pressure gage, applying barely enough pressure to lift the weight. Explain why it would be impossible to calculate the system's efficiency if more pressure than is minimally necessary to move the weight were applied.

4. In terms of work, describe what a pneumatic cylinder does, especially when operated such as in this lab activity.

HYDRAULIC PRESS **L18**

Student Name(s) _____

_____

_____

Hydraulic Press

Purpose

To understand how a hydraulic press produces mechanical advantage.

Laboratory Objectives

- Calculate the ideal mechanical advantage of a hydraulic press by comparing cylinder areas.
- Calculate the actual mechanical advantage of a hydraulic press by comparing output and input forces.
- Verify Pascal's Principle as related to a hydraulic press by calculating pressure in each cylinder.
- Calculate the efficiency of the hydraulic press by comparing output work to input work.
- Calculate the efficiency of the hydraulic press by comparing AMA to IMA.
- Explain and describe Pascal's Principle as it applies to a hydraulic press.
- Explain and describe how a hydraulic press produces mechanical advantage.

Main Ideas

- In a closed cylinder, work is done when the fluid under pressure is displaced.
- Pascal's Principle states that the pressure in a fluid is the same throughout the fluid and in all directions in the fluid.
- According to Pascal's Principle, although the cylinders have different areas and forces, the pressure in the two cylinders is the same.
- A hydraulic press is a system of two connected cylinders of different diameter which produce a mechanical advantage of forces.
- The ideal mechanical advantage of a hydraulic press is the ratio of cylinder areas.
- The actual mechanical advantage of a hydraulic press is the ratio of output and input forces.

- The efficiency of a hydraulic press is the ratio of the output and input work or the ratio of the actual mechanical advantage and the ideal mechanical advantage.

Discussion

In one form or another, devices that create a mechanical advantage are used every day. They may not be identified as such, but lifting an automobile in a service station, pressing a brake pedal, or stamping out sheet metal parts are all examples of mechanical advantage being created in fluid systems. These all use the same physical principle – the hydraulic press – to get more force out than that put in. The hydraulic press is a system where two cylinders (pistons) of different sizes are connected together. A small force applied to the shaft rod of the smaller cylinder will produce a larger force on the shaft rod of the larger cylinder. *Pascal's Principle* states that the fluid pressure is the same throughout a fluid system and that the pressure acts in all directions. This is an important aspect of the hydraulic press: fluid pressure is the same in both cylinders, but with different cross-sectional areas, the cylinders produce different forces.

The cross-sectional area of the cylinders is calculated using $A = \pi r^2$ or $A = \dfrac{\pi d^2}{4}$.

A schematic for the hydraulic press is shown at right. The output cylinder is the large cylinder at right with output cylinder area A_{out}, output force F_{out}, and output rod (stroke) distance d_{out}.

The small input cylinder is much the same with input cylinder area A_{in}, input force F_{in}, and input rod (stroke) distance d_{in}.

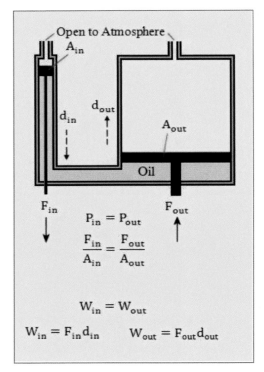

The cylinders and the space between the cylinders is filled with an incompressible fluid, here mineral oil.

Pascal's Principle applies in that the pressure throughout the fluid is the same throughout the system and acts in all directions. This means the pressure in the small cylinder equals the pressure in the large cylinder.

$$p_{in} = p_{out}$$

Since pressure equals force per unit area we have

$$\frac{F_{in}}{A_{in}} = \frac{F_{out}}{A_{out}}$$

This proportion can be rewritten as a proportion comparing areas and forces.

$$\frac{A_{out}}{A_{in}} = \frac{F_{out}}{F_{in}}$$

Note that the output force is larger than the input force by the same factor as the cylinder area comparison ratio, that is, the greater the difference in cylinder size, the greater the output force. The amount by which the input force is multiplied is called the **mechanical advantage**. The **ideal mechanical advantage** (IMA) does not take into account losses, like friction, in the system and is calculated by comparing the areas of the two cylinders. (Careful! The diameters are squared.)

$$IMA = \frac{A_{out}}{A_{in}} = \frac{d_{out}^2}{d_{in}^2}$$

Of course there are frictional losses in the hydraulic press. When the actual forces are measured, we can calculate the **actual mechanical advantage** (AMA) by comparing the output force to the input force.

$$AMA = \frac{F_{out}}{F_{in}}$$

Note that the **input work** and **output work** the hydraulic press is much like the pulley systems we studied in that a small input force moves through a long distance to produce a large output force moving through a small distance. Efficiency can also be calculated in the same way as in the pulley systems.

$$W_{in} = F_{in}d_{in} \quad \text{and} \quad W_{out} = F_{out}d_{out}$$

As with all systems, the **efficiency** can be calculated by comparing output work to input work.

$$\eta = \frac{W_{out}}{W_{in}}$$

Efficiency can also be calculated by comparing the AMA to the IMA.

$$\eta = \frac{AMA}{IMA}$$

Equipment

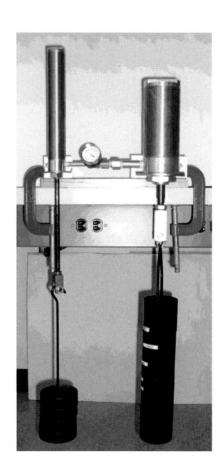

Hydraulic press apparatus consisting of one large bore cylinder and one small bore cylinder filled with mineral oil.

Pressure gage in the line connecting the two cylinders.

Two heavy-duty weight sets.

Two C-clamps.

Ruler or meter stick.

Procedures

1. Clamp the support board and press apparatus to the lab bench.

2. Record the value of the diameters of the cylinders (pistons) in the data table. Calculate the cross-sectional area of the cylinders and record this value in the data table. (Check with instructor if the cylinders are not clearly labeled.)

3. Using $\frac{A_{out}}{A_{in}}$, calculate the Ideal Mechanical Advantage (IMA) of the hydraulic press. Show this in the table provided below.

4. On the clevis of the large cylinder, hang as much weight as is possible using the heavy duty weight set. Record this value in the table as F_{out}. (The cylinder rod should fully extend downward while the small cylinder rod will move up.)

5. Using the IMA value from step 3 in $\frac{F_{out}}{F_{in}}$, calculate the input force F_{in} that theoretically should lift F_{out}. Hang this weight to the small cylinder rod clevis.

6. Continue to add weight a bit at a time until the systems moves and the large cylinder lifts its weight. Record this value in the data table as F_{in}.

7. Remove the F_{in} weight and allow the rods to shift back to their original positions. Hang the F_{in} weight again but this time measuring the distances the rods move, both the small (d_{in}) and large (d_{out}) cylinders. The small cylinder rod should move further than the large cylinder rod. Record these distances in the table provided.

8. Using $AMA = \frac{F_{out}}{F_{in}}$, calculate the Actual Mechanical Advantage of the system, recording this in the table provided.

9. Using $\eta = \frac{AMA}{IMA}$, calculate the efficiency of the system.

10. Calculate the input and output work using W=Fd. Record these values in the table provided.

11. Using $\eta = \frac{W_{out}}{W_{in}}$, calculate the efficiency.

Data

Cylinder Areas:

Small Piston Diameter d_{in} = _____ in   Small Piston Area A_{in} = _____ in^2

Large Piston Diameter d_{out} = _____ in   Large Piston Area A_{out} = _____ in^2

Ideal Mechanical Advantage:

$$IMA = \frac{A_{out}}{A_{in}} = \text{_____} = \text{_____}$$

Weights:

F_{out} = _____ F_{in} = _____

Distances:

d_{out} = _____ d_{in} = _____

Actual Mechanical Advantage:

$$AMA = \frac{F_{out}}{F_{in}} = \text{_____} = \text{_____}$$

Efficiency by Mechanical Advantage:

$$\eta = \frac{AMA}{IMA} = \text{_____} = \text{_____}$$

Work:

$$W_{in} = F_{in}\, d_{in} = \text{_____}$$

$$W_{out} = F_{out}\, d_{out} = \text{_____}$$

Efficiency by Out/In Work Comparison:

$$\eta = \frac{W_{out}}{W_{in}} = \text{_____} = \text{_____}$$

Wrap-Up

1. Explain what is meant by Pascal's Principle. Describe how Pascal's Principle applies to the hydraulic press.

2. Explain the difference between IMA and AMA.

3. In terms of work, explain what a hydraulic press does.

4. You calculated efficiency in two ways, one comparing AMA to IMA and another comparing the input and output work. Explain why these should agree.

5. Why is the AMA always less than the IMA?

Student Challenge

Calculate the IMA when a one inch diameter cylinder is coupled to a 10 inch diameter cylinder.
In such a system, what force could ideally be produced at the output when a one pound force is applied at the input?

Student Name(s) _____

_____

_____

Work Done by an Electric Gear-Motor

Objectives

- In terms of work, explain what an electric motor does.
- Use an electric motor to lift a load.
- Measure voltage across and current through an electric motor.
- Calculate the work used by an electric motor.
- Calculate the mechanical work produced by an electric motor.
- Using the work ratio, determine the efficiency of an electric motor.

Discussion

Electric motors convert electric input work to mechanical output work.

The unifying principle for work is the force-like quantity multiplied by the displacement. In electrical systems the force-like quantity is the voltage (V) measured in volts and the displacement is the charge (q) measured in coulombs. Since there is no such thing as a coulomb meter, the student must measure the current (I) in amperes.

Electric current is the rate at which charge is moved or current = charge/time.

Expressed as a formula current is $I = \dfrac{q}{t}$ where 1 ampere equals 1 coulomb/second.

Manipulating this formula for q we have $q = It$. As a result, It can be substituted into the electrical work formula so that it contains three quantities that can be measured; the electrical work is the product of the voltage, current, and time.

$$W = Eq$$

$$W_{in} = EIt$$

where V is the voltage in volts and I is the current in amperes and t is in seconds. One Volt-Amp-Sec = one Joule.

In this lab the output work is linear mechanical, a lifted weight. Applying the unifying principle of work to linear mechanical systems we have

$$W_{out} = Fd$$

where F the force (or weight) in Newtons, d is the distance the weight is lifted.

The efficiency of this system is the ratio of the output power to the input power or

$$\eta = \frac{W_{out}}{W_{in}}$$

Equipment

DC gear-motor w/ pulley (take-up drum)
DC power supply & banana leads
Heavy duty weights & string
Two multi-meters w/leads (or use on board meters)
Stop watch, timer, or ordinary watch
Heavy duty support stand

Procedure

1. Secure the take-up drum or pulley to the motor shaft and attach the motor to the heavy duty support stand.

2. Securely attach a cord to the take-up drum. Hang weight from take-up drum on the string. Use the given weight for each trial:
   Trial 1: 2kg = 19.6 N
   Trial 2: 4kg = 39.2 N
   Trial 3: 6kg = 58.9 N
   Trial 4: 8 kg = 78.5 N
   Remember the weight here is the force where w = mg.

3. Connect the motor to the power supply. (It can turn in either direction depending on polarity.)

4. Connect the voltmeter across the motor in parallel. Connect the ammeter in series with the motor. (An option here is to use the power supply on-board meters. There are power supplies available, such as the one shown above, that have built-in meters, eliminating the need for separate meters, simplifying the set-up.)

5. Turn on the power supply, setting it to 12 volts, the operating voltage of the motor. Check overall operation of the system including the meters with a dry run or two. (It might be necessary to switch the polarity of the voltage to the motor to let the weight back down, or to turn the motor the other direction. Otherwise simply unwrap the cord to set the weight back down.)

Data and Calculations Table

Trial	F (N)	d (m)	W_{out} (J)	E (volts)	I (amps)	t (sec)	W_{in} (J)	η
1	19.6			12				
2	39.2			12				
3	58.9			12				
4	78.5			12				

Wrap-Up

1. What is the unifying equation for *work*?

2. How does the unifying equation for work apply to electrical systems?

3. In terms of work, explain what an electric motor does.

4. Explain why it is necessary to use units of Newtons rather than kilograms for weight measure in the above calculations.

5. Why are metric units used exclusively when performing calculations involving electrical systems?

6. The gear system creates a mechanical advantage and slows the speed, but how do you suppose the gear system affects the efficiency? Would the efficiency be better or worse if the gears weren't used? Explain your answers.

7. The efficiency of the system might be very low. Why are these devices still utilized when their efficiencies tend to be so poor?

8. As more weight was added, there was a change in the efficiency. According to your data, is there any pattern in this change of efficiency with changes in load weight?

9. From your data table, predict the efficiency when a load of 12 kg is used. (Optional: The students might do an extra trial in order to test their prediction.)

Student Name(s)_____

_____

_____

Work Done by a Water Pump

Lab Objective

When you've finished this lab, you should be able to do the following:

- Determine the pump work required to lift a given volume of water to a given height using the fluid work formula, **W =pV**.
- Determine the pump work required to lift a given volume of water to a given height using the mechanical equivalent of fluid work, **W = Fd**.
- Determine the electrical work "consumed" by the pump using the formula **W = EIt**.
- Determine the efficiency of a water pump by comparing the fluid work to the electrical work using the formula **Efficiency = Output Work ÷ Input Work**, expressing the efficiency as a percent.

Main Ideas

- Work must be done to lift liquids to a higher level.
- The work done must overcome the force of gravity (weight) acting on the liquid.
- The pressure in a pipe filled with liquid is equal to the weight density of the liquid times the vertical height of the column of fluid, $p = \rho h$.
- The efficiency of a pump can be calculated from the ratio Work Out / Work In, where Work Out is the fluid work performed and Work In is the electrical work "consumed".
- In terms of work, an electrical water pump converts electrical work into fluid work.

Work done on liquids, like water or oil, often involves the movement of liquids through pipes or hoses. When pipes are horizontal, work must be done to overcome the friction in the pipes to push the liquid through the pipes. When pipes vary in elevation, additional work must be done to overcome the weight of the liquid. In this lab, we'll study the work done by a water pump to lift a volume of water to a particular height and compare this to the electrical work used. The work done to overcome friction in the pipes (fluid drag) is very small here in this lab because the hoses are short. Therefore, the effects of pipe friction will be ignored in the calculation of efficiency.

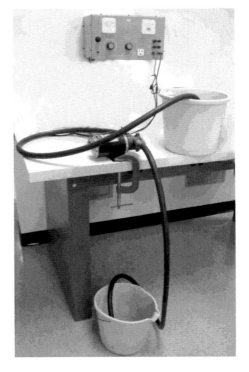

Equipment

12-volt DC RV water pump w/base and connecting hoses

DC Power supply

Two pails, 1-5 gallon

C-clamp

Stop watch

Meter stick or tape measure

Masking tape

Discussion

In terms of work, the electric water pump converts electrical work into fluid work. According to the unifying principle for work, work is calculated by multiplying the force-like quantity to the displacement quantity. The input and output work of the pump is calculated using this principle.

The Input Work is electrical. In electrical systems, the force-like quantity is the voltage, symbolized with either the letter V or E. (Here E will be used because V will be reserved for fluid volume.) The electrical displacement quantity is the charge moved measured in coulombs, symbolized by the letter q. According to the unifying principle, electrical work is then expressed by the formula $W = Eq$. Since there is no such thing as a coulomb meter, the displacement is expressed as the current multiplied by the time, a manipulation of the formula for current, $I = q/t$, where I represents the current in amperes, q represents the charge in coulombs, and t represents the time in seconds. Manipulating the formula for q, we have $q = It$. As a result, we have a useful formula by which electrical work can be calculated using measurable quantities:

Electrical work: $\mathbf{W = EIt}$

The Output Work is fluid. Applying the unifying principle, the force-like quantity is the pressure and the displacement is the volume of fluid moved. Here the pressure is a function of the height the fluid is pumped because the greater the height, the more weight must be overcome. The pressure is calculated using the "stack" formula $P = \rho h$ where ρ is the weight density of water ($\rho = 62.4$ lb/ft^3).

The volume is the amount of fluid displaced measured in cubic feet. Gallons can be converted to cubic feet using 7.48 gallons = 1 ft^3.

Fluid Work: **W = pV**

An alternative method for calculating fluid work is known as "the mechanical equivalent for fluid work". Here the fluid work can be calculated by using W = Fd, where F is the weight of the fluid displaced and d is the height the fluid is pumped.

Mechanical equivalent of fluid work: **W =Fd**

The efficiency of the system is the ratio of the output work to the input work expressed as a percent. The units of measurement for both must be identical so in this case electrical work measured in joules must be converted to ft-lb to match the fluid work units. This is done using the conversion factor 0.738 ft-lb = 1 joule.

Efficiency = Work Out/ Work In

Procedures

Part 1: Set-up

1. Apply a drop of oil to the input port of the water pump.

2. Connect the input and output hoses to the water pump.

3. Secure the water pump to the edge of the table with a c-clamp.

4. Pour one gallon into the bottom container. Mark the level of the water with a pencil or piece of tape.

5. Pour a second gallon into the bottom container. Mark the level of the water with a pencil mark or piece of tape.

6. Pour a third gallon into the bottom container and marking the level of the water. (This will allow for some of the water to be used to prime the pump and allow for more accurate timings.

7. Place the empty container up on the work bench or table.

8. Connect the DC power supply to the pump.

9. Arrange the containers, hoses pump, and power supply as shown in the figure, making sure the input hose is connected to the input port of the pump and that the power supply polarity is correct.

10. Measure the (average) height the fluid is pumped. Record this value in the data table.

Part 2: Collecting Data

11. First read steps 2-8. Before starting to take data, decide who'll operate the power supply, who'll operate the stop watch, and who'll handle the hoses.

12. Prime the input hose by pouring some water into the hose.

13. Turn the power supply on. Adjust the voltage to the rated voltage of the pump.

14. Record the operating voltage of the pump.

15. Record the current drawn by the pump.

16. Record the amount of time the pump takes to move one gallon of water, starting and stopping the stopwatch between markings on the pail. Record this value in the data table.

17. Turn the power supply off.

Data

Height water is lifted: h = _____ ft   Gallons lifted: V = _____ gal

Voltage: E = _____ volts DC   Current: I = _____ amps

Time: t = _____ seconds

Calculations

Input Electrical Work: $W = EIt$

$$W = (\underline{\hspace{1.5cm}}\text{volts})(\underline{\hspace{1.2cm}}\text{amps})(\underline{\hspace{1.2cm}}\text{sec}) = \underline{\hspace{2cm}}\text{ J}$$

$$W = (\underline{\hspace{2.5cm}}\text{ J})\left(\frac{0.738\ \text{ft - lb}}{1\ \text{J}}\right) = \underline{\hspace{2cm}}\text{ lb-ft}$$

Output Fluid Work: $W = pV$

Pressure: $p = \rho h$

$$p = (62.4\ \text{lb}/\text{ft}^3)(\underline{\hspace{2cm}}\text{ ft}) = \underline{\hspace{2cm}}\ \text{lb}/\text{ft}^2$$

Volume:

$$V = (\underline{\hspace{1.5cm}}\text{ gal})\left(\frac{1\,ft^3}{7.48gal}\right) = \underline{\hspace{2cm}}\ ft^3$$

Fluid Work: $W = PV$

$$W = (\underline{\hspace{2.5cm}}\ \text{lb}/\text{ft}^2)(\underline{\hspace{2cm}}\ ft^3) = \underline{\hspace{1.5cm}}\ \text{lb-ft}$$

Output Mechanical Equivalent of Fluid Work: $W = Fd$

The force is the weight of the displaced volume of fluid in pounds.

$$F = (\underline{\hspace{1.5cm}}\text{ gal})(8.34\ \text{lb}/\text{gal}) = \underline{\hspace{2cm}}\text{ lb}$$

$$d = h = \underline{\hspace{1.5cm}}\text{ ft}$$

$$W = Fd = (\underline{\hspace{2cm}}\text{ lb})(\underline{\hspace{2cm}}\text{ ft}) = \underline{\hspace{2cm}}\text{ lb-ft}$$

Efficiency

$$Eff = \frac{\text{Output Work}}{\text{Input Work}} = \underline{\hspace{3cm}} = \underline{\hspace{2cm}}\ \%$$

Wrap-Up

1. What is the unifying equation for work?

2. How does the unifying equation for work apply to pumps?

3. If the pump had to lift the same amount of water to twice the height you measured in this exercise, what changes would you expect in the work calculations and efficiency?

4. You calculated the fluid work using two methods. Should the results of the two methods disagree somewhat, which method is more reliable? Why?

5. Pumps tend to have low efficiencies. Why do we still use them to pump fluids if they're so inefficient?

STUDENT CHALLENGE

Suppose the fluid you pumped in this lab was not water but fuel. What differences would you expect with the pressure, time, and efficiency?

Summary of Special Subunit Work & Efficiency Applications

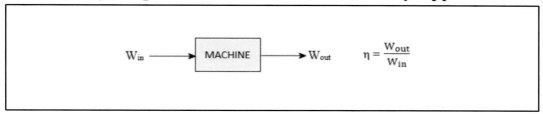

$$W_{in} \longrightarrow \boxed{MACHINE} \longrightarrow W_{out} \qquad \eta = \frac{W_{out}}{W_{in}}$$

$$W_{in} = F_{in}\, d_{in} \longrightarrow \boxed{PULLEY} \longrightarrow W_{out} = F_{out}\, d_{out}$$
$$AMA = \frac{F_{out}}{F_{in}}$$
$$IMA = \frac{d_{in}}{d_{out}}$$
$$\eta = \frac{W_{out}}{W_{in}}$$

$$W_{in} = \tau\theta \longrightarrow \boxed{WINCH} \longrightarrow W_{out} = F_{out}\, d_{out}$$
$$\tau = Fl$$
$$1 \text{ rev} = 2\pi \text{ rad}$$
$$AMA = \frac{F_{out}}{F_{in}}$$
$$\eta = \frac{W_{out}}{W_{in}}$$

$$W_{in} = pV \longrightarrow \boxed{PISTON} \longrightarrow W_{out} = Fd$$
$$p = F/A \qquad A = \pi r^2$$
$$V = \pi r^2 h$$
$$\eta = \frac{W_{out}}{W_{in}}$$

$$W_{in} = F_{in}\, d_{in} \longrightarrow \boxed{PRESS} \longrightarrow W_{out} = F_{out}\, d_{out}$$
$$p_1 = p_2$$
$$\frac{F_1}{A_1} = \frac{F_2}{A_2}$$
$$p = F/A \qquad A = \pi r^2$$
$$AMA = \frac{F_{out}}{F_{in}} \qquad IMA = \frac{A_2}{A_1}$$

$$W_{in} = EIt \longrightarrow \boxed{MOTOR} \longrightarrow W_{out} = \tau\theta \ (\text{or } Fd) \qquad \eta = \frac{W_{out}}{W_{in}}$$

$$W_{in} = EIt \longrightarrow \boxed{PUMP} \longrightarrow W_{out} = pV \ (\text{or } Fd) \qquad \eta = \frac{W_{out}}{W_{in}}$$
$$p = \rho_w h$$

Example 4-27

A pulley system is used to lift a 450 pound weight to a height of 4 feet. The worker must pull 36 feet of line with a force of 60 pounds to raise the load the 4 feet. Calculate (**a**) the actual mechanical advantage and (**b**) the efficiency.

Solution:

a. $AMA = \dfrac{F_{out}}{F_{in}} = \dfrac{450\,\cancel{lb}}{60\,\cancel{lb}} = 7.5$

b. $\eta = \dfrac{W_{out}}{W_{in}} = \dfrac{F_{out}\,d_{out}}{F_{in}\,d_{in}} = \dfrac{450\cancel{lb}(4ft)}{60\cancel{lb}(36ft)} = \dfrac{1800}{2160} = 0.833 = 83.3\%$

Example 4-28

Calculate the force generated by a three inch diameter piston operating at 90psi.

Solution:

$F = pA = p(\pi r^2) = 90lb/\cancel{in}^2\,(3.14)(1.5\cancel{in})^2 = 636\,lb$

Note how when pressure is multiplied by area, the area units cancel leaving units of force.

Example 4-29

A Pascal press operates with the use of two different sized cylinders. The small input cylinder has a 2 inch diameter and the large output cylinder has a 12 inch diameter. A force of 10 pounds is applied to the input cylinder. Calculate (**a**) the ideal mechanical advantage and (**b**) the output force (assuming the system works at near 100% efficiency).

Solution:

a. $IMA = \dfrac{A_{out}}{A_{in}} = \dfrac{\cancel{\pi}(6in)^2}{\cancel{\pi}(1in)^2} = 36$

b. $MA = \dfrac{F_{out}}{F_{in}}$; $F_{out} = MA(F_{in}) = 36(10lb) = 360\,lb$

When calculating IMA using the ratio of cylinder diameters, π cancels. As a result, IMA can also be expressed as the ratio of either the radii or diameters squared, $IMA = \dfrac{d_o^2}{d_i^2}$.

Example 4-30

A 12 volt electric motor draws 6 amperes for 1 minute while lifting a 16 kg weight to a height of 4 meters. Calculate the efficiency of the motor.

Solution:

$$W_i = EIt = 12V(6A)(60sec) = 4320 \text{ J}$$

$$W_o = Fd = 16kg(9.81 \text{ m/s}^2)(4m) = 628 \text{ J}$$

$$\eta = \frac{W_{out}}{W_{in}} = \frac{628\cancel{J}}{4320\cancel{J}} = 0.145 = 14.5\%$$

Don't forget to multiply the mass in kilograms to the accelerations due to gravity 9.81 m/s² in order to have force in Newtons. Otherwise Joules (N-m) will not be consistent in the efficiency formula.

Example 4-31

Calculate the efficiency of a 110 volt water pump that draws 15A pumping 40 gallons to a height of 60 feet in 16 minutes.

Solution:

$$W_i = EIt = 110V(15A)(960sec) = 1584kJ$$

$$W_o = Fd = 40\cancel{gal}(8.34lb/\cancel{gal})(60ft) = 20,016 \text{ lb-ft}$$

From conversion table, 77.79 lb-ft = 1055 J.

$$20,016 \text{ lb-ft}\left(\frac{1055J}{77.79lb-ft}\right) = 271 \text{ kJ}$$

$$\eta = \frac{W_o}{W_o} = \frac{271 \cancel{kJ}}{1584\cancel{kJ}} = 0.171 = 17.1\%$$

Note how the mechanical equivalent of fluid work was employed here and how the weight of the water in pounds had to be calculated in Fd. Note also how lb-ft was converted to Joules for consistent units in the efficiency formula.

Example 4-32

Calculate the efficiency of a device that consumes 1200 Joules of energy to accomplish 850 Joules of useful work.

Solution:

$$\eta = \frac{W_o}{W_o} = \frac{850 \cancel{J}}{1200\cancel{J}} = 0.708 = 70.8\%$$

Note how terms such as "consumes" and "accomplish" identify the input verses output.

Example 4-33

By inspection, what is the ideal mechanical advantage of the pulley system shown?

Solution:

By counting supporting ropes, and including the rope being pulled on,
$$IMA = 5$$

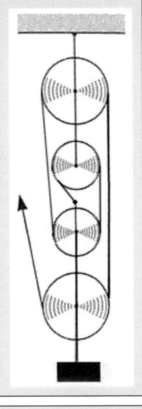

Example 4-34

For the pulley system in Example 4-33, how much weight can ideally be lifted with an input force of 250 Newtons?

Solution:

$$MA = \frac{F_{out}}{F_{in}}$$

$$F_{out} = MA(F_{in}) = 5(250N) = 1250 \ N$$

Example 4-35

For the pulley system in Example 4-33, more force appears at the output than at the input, but this increase in force comes at a price. What length of rope must the operator pull out in order to raise the load one foot?

Solution:

She must pull five times more or
$$d_{in} = 5 \text{ feet of rope.}$$

4.6 STUDENT EXERCISES   WORK & EFFICIENCY APPLICATIONS

1. What is the unifying equation for work?

2. **a.** The units of measurement for work in the English system is _____.

    b. The units of measurement for work in the metric system is _____.

3. The efficiency of a machine is known to be 28%. If 45kJ of work must be accomplished by the machine, how much electrical work must be put in?

4. A block and tackle system has a mechanical advantage of six. When using this 80% efficient system, how much force must be applied to move a 120 pound load?

5. In # 4 above, more force appears at the output than at the input.
    a. Explain how although there is more output force, there cannot be any more work at the output than work put in.

    b. Explain what must be sacrificed, or compensated, at the input to gain the machincal advantage in force.

6. A 110 volt electric motor draws 10 amperes for one minute while lifting a 330N weight to a height of 20 meters. Find
   a. the electrical work "consumed" by the motor.
   b. the work done to raise the load.
   c. the efficiency of the motor.

7. Explain how a Lear jet and a lawn mower will do the same amount of work carrying a person from New York to California.

8. The two pulley systems shown are used to raise a 240 pound weight. For each pulley system, find
   a. the ideal mechanical advantage.
   b. the required ideal input force.

9. A 110V electric water pump draws 15 amps for one hour. At 15% efficiency, how much fluid will be displaced when operating at 490Pa?

10. A two inch diameter hydraulic cylinder has a six inch stroke (shaft movement).
    a. How much force will the cylinder ideally produce at 80psi and full stroke?
    b. How much mechanical work will it do?

11. Explain what is meant by the mechanical advantage of a system.

12. The mechanical actual advantage of a simple chain hoist is 30. Using this system, how much input force is required to lift a 1200 pound load?

13. Calculate the efficiency of a 110V water pump that draws 13A for 28 minutes pumps 100 gallons of water from a well 80 feet deep.

14. Calculate the efficiency of a system that uses 1.24kJ of energy to accomplish 445J of work.

Answers: 1. *Work is the mover quantity multiplied by the displacement quantity, except for thermal systems* **2a.** pound-foot **2b.** Joule **3.** 161kJ **4.** 25 lb **5a.** *There can never be more work out than in because energy cannot be created and also there are always losses in the system. Force is only part of the work equation.* **5b.** *More force can appear at the output because distance is sacrificed at the input.* **6a.** 66kJ **6b.** 6.60kJ **6c.** 10% **7.** *Both carry the same weight the same distance. Time doesn't matter.* **8a(left)** 4 **8a(right)** 3 **8b(left)** 60 lb **8b(right)** 80 lb **9.** 1820 m^3 **10a.** 251 lb **10b.** 126 lb-ft **11.** *The ratio of the output force to the input force.* **12.** ideally 40 lb **13.** 37.7% **14.** 35.9%

Chapter 5   Power

In the chapters leading up to now, we studied several unifying ideas that apply to all of the energy systems. Some of these ideas, such as the unifying principle for work, are formulae-building concepts that apply to the various energy systems, thereby "unifying" the energy systems. In this book, all of these principles have been presented in an order that leads us to this last chapter on a unifying principle for power.

In review, we learned that each energy system has a displacement quantity and a mover quantity. We then combined these two quantities in the unifying principle for work. We also studied the unifying principle of rate, where the displacement quantity is divided by time. Building on these, we are now prepared to understand the unifying principle for **power**, the rate of doing work. If work is to be done more quickly, more power is needed. Capital letter "P" is used to represent power in formulae.

$$P = \frac{W}{t}$$

Energy System	Displacement	Rate	Mover	Work = Mover times Displacement	Power = Work ÷ time	Power = Mover times rate
Translational Mechanical	Distance (d)	$v = d/t$	Force (F)	$W = Fd$	$P = \dfrac{Fd}{t}$	$P = Fv$
Rotational Mechanical	Angle (θ)	$\omega = \theta/t$	Torque (τ)	$W = \tau\theta$	$P = \dfrac{\tau\theta}{t}$	$P = \tau\omega$
Fluid	Volume (V)	$\dot{V} = V/t$	Pressure (p)	$W = pV$	$P = \dfrac{pV}{t}$	$P = p\dot{V}$
Electrical	Charge (q)	$I = q/t$	Voltage (E)	$W = Eq$	$P = \dfrac{Eq}{t}$	$P = EI$
Thermal	Heat (Q)	$\dot{Q} = Q/t$	Temperature Difference (ΔT)	Does not apply. Instead $W = Q$	$P = \dot{Q}$	$P = \dot{Q}$

Applying the unifying principle for power, P = W/t, all of the work formulae are divided by time, "t", to produce the various power formulae.

But there is another way of thinking of power, what might be called a second unifying power principle. Note there are three quantities that comprise power: a mover, a displacement, and time. Grouping the mover and displacement together in the numerator produces the "work over time" principle. But note that if we group the displacement quantity in the numerator with the time in the denominator, we have the equation for rate in that system. For example, note in the above table that the velocity formula d/t appears in the mechanical P = Fd/t formula. As a result, we can think of power as the force times the velocity or P = Fv. In general, power in any particular energy system is the energy system's mover times the energy system's rate. This results in power equals mover time rate. Typically, technicians tend to think of power in these terms, a mover times a rate, because these are the two quantities that are varied and controlled in a system, sacrificing one for the other, with total available power always limited. For example, when you shift gears on a pedal bicycle, and with limited power in your legs, you might sacrifice speed for force when going uphill, but switch gears to get more speed when going downhill. Similarly, a tractor having a limited amount of power might rotate a power take-off shaft at high speed but low torque for a particular implement, say a mower, but rotate the shaft at a slower speed with high torque for another implement, say a tiller or posthole digger. With a limited amount of power, the mover is sacrificed for rate or visa-versa to best serve the situation. So power can be seen as a combination of some type of mover quantity and a rate, with one of these quantities sacrificed for the other depending on the situation.

The units for power are units of work divided by time. In the English system, the units become pound-feet per second or lb-ft/sec. Remember how a distinction was made between lb-ft and ft-lb so that units of work would not be confused with units of torque? But here, with seconds in the denominator, there is no ambiguity and no doubt that power is the measured quantity, so units for power are most often written ft-lb/sec.

In SI, we have N-m/sec. But remember that to distinguish units of work from units of torque, the Joule is used as the unit of work, or 1 N-m = 1J. So in the metric system, we have Joules per second or J/sec for units of power. This combination of units is called the **Watt**.

> 1 joule/sec = 1 watt

James Watt

James Watt did *not* invent the steam engine and he did *not* invent the watt unit of power. What he *did* do was improve the steam engine to the point that the machine would work reliably and efficiently. He also developed rotating shaft systems, what we now call "power take-offs" so that the power produced by the engine can be applied to many different types of applications, including the electrical generators developed by Nikola Tesla and Thomas Edison.

The Chicago World's Columbian Exposition of 1893, better known as the Chicago World's Fair, truly amazed people. Back then, people were not familiar with electricity and used animals for their source of power. Most of the fair's exhibits were dedicated to the new electrical inventions, especially Tesla's amazing developments in the production, distribution, and applications of alternating current. People had never seen electricity like this before: the lights, fountains, Ferris wheels, and so much more, all electrically driven, *and all powered by generators rotated by Watt's steam engines*. It is because of Watt's reliable steam engines rotating the electrical generators that the joule per second was named after him, the watt.

Horsepower

Initially Watt had difficulty marketing his machine. Foot-pounds per second didn't mean much to people who were only familiar with the power provided by animals. So he studied the power of small donkeys used in mining operations, measuring their steady work rate (power) over lengthy periods of time. Given that different animals of different size and temperament vary greatly in the power they can provide, he was able to come up with an approximate value for the power of a typical draft animal. He called it the **horsepower**, abbreviated "hp". He decided a typical draft animal can raise a 550 pound weight one foot in one second. People better understood the tremendous and consistent amounts of power produced by his machines when he described the machines' power in terms of "horsepower". Although draft animals are seldom used today, the popularity of the horsepower standard remains, and the unit is still commonly used. The "size" of equipment such as pumps and motors are often expressed in horsepower.

$$1 \text{ hp} = 550 \text{ ft-lb/sec}$$

Caution! Horsepower is never used in a power calculation. Horsepower is never plugged into a power formulae, and no power formula will produce horsepower units. All power formulae are based on the ft-lb/sec, watt, and as we shall soon see, thermal rate units.

James Watt (1736 – 1819)

James Watt was a Scottish inventor and mechanical engineer whose improvements to the Newcomen steam engine were fundamental to the changes brought by the Industrial Revolution in both his native Great Britain and the rest of the world.

While working as an instrument maker at the University of Glasgow, Watt became interested in the technology of steam engines. He realised that contemporary engine designs wasted a great deal of energy by repeatedly cooling and re-heating the cylinder. Watt introduced a design enhancement, the separate condenser, which avoided this waste of energy and radically improved the power, efficiency, and cost-effectiveness of steam engines. Eventually he adapted his engine to produce rotary motion, greatly broadening its use beyond pumping water.

He developed the standard of horsepower. The SI unit of power, the watt, was named after him.

Thermal Units of Power

Our unifying principles do well to unify the energy systems. But as useful as these unifying principles are, our unifying principles at times do not apply to thermal systems. Remember the exception regarding thermal work, that thermal work is equal to thermal heat or $W = Q$. As a result, the unifying principle for power actually does apply to thermal systems in the sense that power is work divided by time. As a result, thermal power is equal to thermal rate or $P = Q/t = \dot{Q}$.

The units of measurement for thermal power are then equal to units of thermal rate, heat over time, or Btu/hr and cal/sec. We studied thermal rate in the chapter on rate. Units of thermal rate can be converted directly to any other units of power.

Power Conversion Factors

There are five useful units of power: horsepower, ft-lb/sec, watt, Btu/hr, and cal/sec. But a conversion table that would provide a conversion factor converting any one of these directly to any other would be quite lengthy. So to simplify things, technicians use only a short and tidy four factors, each equal to one horsepower. Once in horsepower, power can be converted to any of the others.

1 hp = 550 ft-lb/sec
1 hp = 746 W
1 hp = 2545 Btu/hr
1 hp = 178.2 cal/sec

5.1   Power in Translational Mechanical Systems

In translational mechanical systems, work is done when an object is moved over a distance. The rate at which this work is done is the translational power.

Here we will limit discussion to the power needed to raise a weight to a particular height within a particular time.

$$P = \frac{Fd}{t} = Fv$$

Example 5-1

How much horsepower is needed to lift a 300 pound weight to a height of 20 feet in half a minute?

Solution:

$$P = \frac{Fd}{t} = \frac{(300 \text{ lb})(20 \text{ ft})}{30 \text{ sec}} = 200 \text{ ft-lb/sec}$$

$$P = 200 \text{ft-lb/sec}\left(\frac{1\text{hp}}{550 \text{ ft-lb/sec}}\right) = 0.364 \text{ hp}$$

Example 5-2

How much horsepower is needed to lift 100 kilograms to a height of 5 meters in half a minute?

Solution:

$$F = (100 \text{ kg})(9.81 \text{ m/s}^2) = 981 \text{ N}$$

$$P = \frac{Fd}{t} = \frac{(981 \text{ N})(5 \text{ m})}{30 \text{ sec}} = 164 \text{ W}$$

$$P = 164\text{W}\left(\frac{1\text{hp}}{746\text{W}}\right) = 0.220 \text{ hp}$$

Example 5-3

How much horsepower is needed to lift 1000 pounds at a speed of 3 ft/sec?

Solution:

$$P = Fv = (1000 \text{ lb})(3 \text{ ft/sec}) = 3000 \text{ ft-lb/sec}$$
$$P = 3000\text{ft-lb/sec}\left(\frac{1\text{hp}}{550 \text{ ft–lb/sec}}\right) = 5.45 \text{ hp}$$

Example 5-4

An electric motor can produce 4000 watts of output power. At what speed will the motor lift a 1600 N load?

Solution:

$$P = Fv \text{ (manipulating the formula for v)}$$
$$v = \frac{P}{F} = \frac{4000\text{W}}{1600\text{N}} = 2.5 \text{ m/s}$$

Example 5-5

How much horsepower is needed to lift 1000 kg at a speed of 2 m/sec?

Solution:

$$P = Fv = (1000 \text{ kg})(9.81\text{m/s}^2)(2 \text{ m/sec}) = 19,620 \text{ W}$$

$$P = 19,620\text{W}\left(\frac{1\text{hp}}{746\text{W}}\right) = 26.3 \text{ hp}$$

In mechanical rotational systems, work is the product of the torque and the angular displacement, expressed as $\tau\theta$. Applying our unifying principle for power, work divided by time, we have $P = \dfrac{\tau\theta}{t}$.

Also, by grouping the angle and time together as one quantity, the angular rate, we have $P = \tau\omega$.

Remember that the angle in the formula for rotational work must be in radians in order for the units power to end up in ft-lb/s or watts.

Also, $\tau = Fl$.

Rotational Power
$P = \dfrac{\tau\theta}{t} = \tau\omega$

Example 5-6

A "planer" is a powered woodworking tool. The planer cutting head must operate at 4000 rpm and develops 70 N-m or torque. Calculate the power produced at the cutter head in Watts and horsepower.

Solution:

$$\omega = 4000 \text{ rev/min} = 419 \text{ rad/sec}$$
$$P = \tau\omega = 70 \text{ N-m}(419 \text{ rad/sec}) = 29330 \text{ W}$$
$$P = 29330 \cancel{W}\left(\frac{1 \text{ hp}}{746 \cancel{W}}\right) = 39.3 \text{ hp}$$

Convert rpm to rad/s by multiplying by 6.28 rad/rev and dividing by 60 sec/min.

Example 5-7

A posthole digger operating at 540 rpm is powered by the PTO (power take-off) shaft of a 30 hp compact tractor. Calculate the torque produced by the posthole digger.

Solution:

$$\omega = 540 \text{ rpm} = 56.5 \text{ rad/sec}$$
$$P = 30 \cancel{hp}\left(\frac{550 \text{ ft-lb/sec}}{1 \cancel{hp}}\right) = 16500 \text{ ft-lb/sec}$$
$$P = \tau\omega$$
$$\tau = \frac{P}{\omega} = \frac{16500 \text{ ft-lb/sec}}{56.5 \text{ rad/sec}} = 292 \text{ ft-lb}$$

In fluid systems, work is the product of the pressure and the volume of fluid displaced, expressed as $W = pV$. Applying our unifying principle for power, work divided by time, we have $P = \dfrac{pV}{t}$. Also, by grouping the volume and time together as one quantity, the volume flow rate, we have $P = p\dot{V}$.

Fluid Power

$$P = \frac{pV}{t} = p\dot{V}$$

Example 5-8

650 cfh flows through an air motor while operating at 90 psi. Calculate the power consumed by the motor.

Solution:

$$\dot{V} = 650 \text{ ft}^3/\text{hr} = 0.181 \text{ ft}^3/\text{s}$$
$$p = 90 \text{ psi} = 12960 \text{ lb/ft}^2$$
$$P = p\dot{V} = 12960 \text{ lb/ft2}(0.181 \text{ ft3/sec}) = 2346 \text{ ft-lb/sec}$$

Cfh (ft³/hr) is converted to ft ³/sec by dividing by 3600 sec/hr. Psi is converted to lb/ft² by multiplying by 144.

Example 5-9

Calculate the power (in horsepower) provided by a water pump that can move 1000 gallons of seawater to a height of 60 feet in ten minutes.

Solution:

$$p = 62.4 \text{ lb/ft}^3(60 \text{ ft})(1.04) = 3890 \text{ lb/ft}^2$$

$$\dot{V} = 1000\text{gal}\left(\frac{1\text{ft}^3}{7.48 \text{ gal}}\right) = 134 \text{ ft}^3$$

$$P = p\dot{V} = p\frac{V}{t} = 3890 \text{ lb/ft}^2(134 \text{ ft}^3)/600 = 869 \text{ lb-ft/sec}$$

$$P = 869 \text{ lb-ft/sec}\left(\frac{1\text{hp}}{550 \text{ ft-lb/sec}}\right) = 1.58 \text{ hp}$$

Example 5-10

How much horsepower is required of a water pump that must move 50,000 gallons to a height of 120 feet in two hours?

Solution #1: Using the stack formula.

$$p = \rho_w h = 62.4 \text{ lb/ft}^3(120\text{ft}) = 7490 \text{ lb/ft}^2$$

$$V = 50{,}000\cancel{\text{gal}}\left(\frac{1\text{ft}^3}{7.48 \cancel{\text{gal}}}\right) = 6680 \text{ ft}^3$$

$$t = 2 \text{ hrs} = 7200 \text{ sec}$$

$$P = \frac{pV}{t} = \frac{7490\text{lb/ft}^2(6680\text{ft}^3)}{7200\text{sec}} = 6949 \text{ ft-lb/s}$$

$$P = 6949 \cancel{\text{ft-lb/s}}\left(\frac{1 \text{ hp}}{550 \cancel{\text{ft-lb/s}}}\right) = 12.6 \text{ hp}$$

Solution #2: Using the mechanical equivalent of fluid power.

$$F = 50{,}000 \cancel{\text{gal}}\left(\frac{8.34 \text{ lb}}{1 \cancel{\text{gal}}}\right) = 417{,}000 \text{ lb}$$

$$P = \frac{Fd}{t} = \frac{417{,}000 \text{ lb}(120\text{ft})}{7200\text{sec}} = 6950 \text{ ft-lb/s} = 12.6 \text{ hp}$$

In electrical systems, work is the product of the voltage applied and the charge displaced, expressed as W = Eq. Applying our unifying principle for power, work divided by time, we have $P = \dfrac{Eq}{t}$. Also, by grouping the charge and time together as one quantity, the current, we have P = EI.

Remember that in electrical systems, three quantities are represented in Ohm's Law: the voltage (E), the current (I), and the resistance (R), where E = IR. By substituting Ohm's Law in to P = EI, we can create two more electrical power formulae. This saves us a step having to employ Ohm's Law, with three formulae ready to calculate power given any two of the three quantities in Ohm's Law.

We learned in an earlier chapter that resistors oppose the current flow, with their resistance measured in Ohms (Ω). An important property of resistors is that the physical size of the resistor has nothing to do with the value of its resistance. A very small resistor, one that might fit in a microchip, could have a resistance in the millions of ohms (MΩ), while a resistor a foot long could have only one ohm of resistance. The physical size of the resistor is its power in watts. The larger the resistor, more current can flow through it. Being physically larger, it can dissipate more heat. A half-watt resistor is approximately a half inch long.

Electrical Power

$$P = EI \quad P = I^2R \quad P = E^2/R$$

Example 5-11

How much power is dissipated by a 1kΩ resistor passing 0.5A at 12 volts?

Solution:

$$P = EI = 12V(0.5A) = 6W$$

Example 5-12

How much power is dissipated by a 1kΩ resistor passing 0.75mA?

Solution:

$$P = I^2R = (0.75 \times 10^{-3} A)^2 (1000\Omega) = 0.563mW$$

Example 5-13

How much power is dissipated by a 4.7MΩ resistor at 24V?

Solution:

$$P = E^2/R = \frac{(24V)^2}{4.7 \times 10^6 \Omega} = 0.123mW$$

Example 5-14

What is the maximum current a 0.25W resistor can safely pass at 12 volts?

Solution:

$$P = EI$$
$$I = \frac{P}{E} = \frac{0.25W}{12V} = 20.8 \text{ mA}$$

Watt-hour Meters

When the electric company provides electric power to your home, a meter is placed at the electrical entrance to measure how much energy you use. If we know how much electrical power is used, and also the amount of time the power is used, the electrical energy (technically work in Joules) can be calculated. Remember that P = W/t, so W = Pt.

Normally in physics, to calculate the work or energy, we would multiply the power in watts to the time in seconds. This results in units of work, the joule (1W = 1J/s). But electric power companies do things a bit differently. They measure the power in watts and the time in hours, so when P and t are multiplied the units are watt-hours. This is what the electric company charges us in, cents per watt-hour, or more commonly, cents per kilowatt-hour.

We learned in earlier chapters that although our unifying principles are useful in unifying four energy systems, thermal systems tend to be an exception. But when it comes to thermal power, the work over time, or energy over time, principle applies, because in thermal systems the energy is the heat itself, measured in Btu's or calories. As a result, in thermal systems, thermal power equals thermal rate.

$$\text{Thermal Power}$$
$$P = \frac{W}{t} = \frac{Q}{t} = \dot{Q}$$

Example 5-15

Calculate the thermal power of an oil-fired water heater (in watts) that can heat 45 gallons of water from 90°F to 150°F in 30 minutes.

Solution:

$$t = 30 \text{ minutes} = 0.5 \text{ hr}$$

$$\text{``m''} = 45\cancel{\text{gal}}\left(\frac{8.34\text{lb}}{1\cancel{\text{gal}}}\right) = 376 \text{ lb}$$

$$\Delta T = 150 - 90 = 60\text{F}°$$

$$Q = mc\Delta T = 376\text{lb}(1 \text{ Btu/lb-F}°)(60\text{F}°) = 22,600 \text{ Btu}$$

$$P = \frac{Q}{t} = \frac{22,600\text{Btu}}{0.5\text{hr}} = 45,200 \text{ Btu/hr}$$

$$P = (45,200\cancel{\text{Btu/hr}})\left(\frac{1\cancel{\text{hp}}}{2545\cancel{\text{Btu/hr}}}\right)\left(\frac{746\text{W}}{1\cancel{\text{hp}}}\right) = 13.2\text{kW}$$

Example 5-16

A heat exchanger is a device that displaces heat. The rate at which the exchanger displaces heat is the exchanger's power.

A particular heat exchanger condenses 12 liters of 100°C steam down to room temperature, 22°C, in three hours.

Calculate the power of the heat exchanger in watts.

Solution:

$t = 3$ hours $= 10,800$ sec

$m = 12kg = 12,000g$ (water 1kg/L)

Latent heat:

$Q_v = mH_v = 12,000g(540cal/g) = 6.48 \times 10^6 cal$

Sensible heat:

$\Delta T = 100 - 22 = 78C°$

$Q = mc\Delta T = 12,000g(1\ cal/g\text{-}C°)(78C°) = 9.36 \times 10^5\ cal$

Power:

$$P = \frac{Q_T}{t} = \frac{6.48 \times 10^6 cal + 9.36 \times 10^5 cal}{10,800sec} = 687\ cal/sec$$

$$P = (687cal)\left(\frac{1hp}{178.2cal}\right)\left(\frac{746W}{1hp}\right) = 2880W$$

5.6 Special Subunit   Power & Efficiency Applications

So far in our study of power, we studied power in each of the individual energy systems. In this section, we'll examine how power is converted from one form to another by various types of machines and devices.

The purpose of most machines is to change power from one form to another. For example, an electric motor coverts electrical power into mechanical power. But all of the power put into the electrical motor (P_{in}) does not all appear at the output (P_{out}) as mechanical power: there are always losses that reduce the efficiency.

We already studied work and efficiency in a previous section. Power efficiency is very similar, with one form of power at the input of the machine and another type of power at the output. And much like work, the efficiency of the machine is the ratio of the output power to the input power. As a result, because the time interval is the same at the input and output, efficiency can be calculated in either way, either the ratio of the work or the ratio of power.

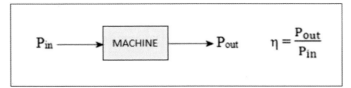

The following examples and exercises can often involve solving two power problems, one at the input and one at the output. These are then divided to determine the efficiency.

There's a second type of problem addressed in this section: sizing the machine. Given enough information about the power that must be accomplished at the output and also the efficiency of commonly available machines, the input power is calculated in order to determine the minimum sized machine that must be installed. The machines are usually rated in horsepower or watts.

Energy System	Power = Work ÷ time	Power = Mover times rate
Translational Mechanical	$P = \dfrac{Fd}{t}$	$P = Fv$
Rotational Mechanical	$P = \dfrac{\tau\theta}{t}$	$P = \tau\omega$
Fluid	$P = \dfrac{pV}{t}$	$P = p\dot{V}$
Electrical	$P = \dfrac{Eq}{t}$	$P = EI$
Thermal	$P = \dot{Q}$	$P = \dot{Q}$

ELECTRIC GEAR MOTOR L21

Student Names(s) _____

Power & Efficiency: Electric Gear-Motor

Objectives

- In terms of power, explain what an electric motor does.
- Use an electric motor to lift a load.
- Measure voltage across and current through an electric motor.
- Calculate the power used by an electric motor.
- Calculate the mechanical power produced by an electric motor.
- Using power ratio, determine the efficiency of an electric motor.

Discussion

Electric motors convert electric input power to mechanical output power.

The unifying principle for power is the rate of doing work or P=W/t. Since work is equal to the force-like quantity multiplied by the displacement, this unifying principle can also be expressed as the force-like quantity multiplied by the rate of that system. Applying this second version of the unifying principle to the electrical input power of the electric motor we have

$$P = EI$$

where E is the voltage in volts and I is the current in amperes. One volt-amp = one watt.

In this lab the output power is linear mechanical, a lifted weight. Applying the primary unifying principle of power, P = W/t, to linear mechanical systems we have

$$P = \frac{Fd}{t}$$

where F the force (or weight) in Newtons, d is the distance the weight is lifted, and t is the time it takes to lift the weight the distance. N-m/sec = J/sec = watt.

The efficiency of this system is the ratio of the output power to the input power or

$$\eta = \frac{P_{out}}{P_{in}}$$

Equipment

DC electric gear-motor w/ pulley or take-up drum

DC power supply

Gear assembly w/belt

Banana leads

Heavy duty slotted weights

String

Two multi-meters w/leads

Stop watch

Heavy duty support stand

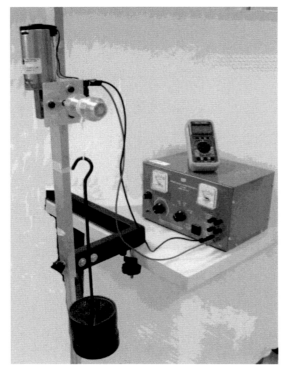

Procedure

1. Secure the take-up drum or pulley to the motor shaft and attach the motor to the heavy duty support stand. (The apparatus can be set up in a variety of other ways, such as by clamping the motor to a lab bench. The idea here is to lift a weight with the motor through the gear system.)

2. Securely attach a cord to the gear assembly output pulley (or take-up drum). Attach the other end to the weights. For Trial One use 3 kilograms. For Trial Two use 6 kilograms. Remember w = mg.

3. Connect the motor to the power supply using two banana leads.

4. Connect the voltmeter across the motor in parallel. Connect the ammeter in series with the motor.

5. Have your instructor look over the setup before continuing.

6. Turn on the power supply, setting it to no more than 12 volts. Check overall operation of the system including the meters with a dry run or two. (It might be necessary to switch the polarity of the voltage to the motor to let the weight back down, or to turn the motor the other direction.)

Data and Calculations Table

Trial	F (N)	d (m)	t (sec)	P$_{out}$ (W)	E (volts)	I (amps)	P$_{in}$ (W)	η
1								
2								

Wrap-up

1. In terms of power, explain what an electric motor does.

2. Explain why it is necessary to use units of Newtons rather than kilograms for weight measure in the above calculations.

3. Explain how both versions of the unifying principle for power (work/time and mover times rate) were applied in this lab in order to calculate input and output power.

4. The gear system creates a mechanical advantage, but how do you suppose the gear system affects the efficiency? Would the efficiency be better or worse if the gears weren't used? Explain your answers.

Student Name (s) _____

_____

_____

Power & Efficiency: Electric Transformer

Purpose

To study the voltage, current, and power relationships of an electrical transformer.

Discussion

The purpose of an electrical transformer is to increase or decrease the voltage from an AC source. Because of losses, this is done at the expense of current. The electrical transformer uses magnetic induction to increase or decrease a voltage. It is made up of an input coil of wire, and output coil of wire, and a soft iron core. The ratio of the number of turns of wire in the input and output coils determines how much the voltage is increased or decreased. If the voltage is increased, the transformer is referred to as a *step-up* transformer. If the voltage is decreased, it is referred to as a *step-down* transformer.

While the transformer voltage is changed between the input to output coils, power is lost during the process. At best (an ideal transformer), the power remains constant. Normally, the system is not perfectly efficient, and considerable power is lost between the primary and secondary coils.

The essential parts of a transformer are pictured below. The input side of the transformer is called the *primary,* while the output side of the transformer is called the **secondary**. In an ideal transformer, the electrical power at the primary equals the electrical power at the secondary or

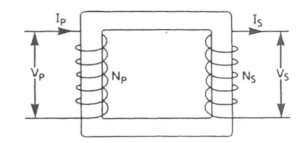

$$P_p = P_s \text{ or } E_p I_p = E_s I_s$$

Since the VI product (power) is not increased, a *voltage increase occurs at the cost of current decrease.*

Of course there is no such thing as an ideal transformer because there are always losses. The efficiency of the transformer is the ratio of the power out at the secondary to the power in at the primary or

$$\eta = \frac{E_s I_s}{E_p I_p}$$

The alternating current through the primary and secondary coil windings creates a changing magnetic field throughout the soft iron core, linking the two coils. The induced voltage in each winding is proportional to the number of turns of wire. This relationship between the number of turns of wire and the induced voltage can this be expressed

$$\frac{E_p}{E_s} = \frac{N_p}{N_s}$$

It follows that when there are more secondary windings, the transformer is in step-up mode, and when there are less secondary windings the transformer is in step-down mode. Most transformers can be connected in either mode, with either coil serving as the input.

The value of the resistor selected to serve as the transformer load will dramatically affect the transformer's efficiency in the two modes of operation, step-up and step-down. Impedance matching is beyond the scope of this discussion/lab.

Equipment

Student transformer assembly

0-16 volt AC power supply

Four digital multi-meters

Resistor, $\approx 1k\Omega$

Two toggle switches (optional)

Banana leads

Patch bay or Perf Board to mount resistor

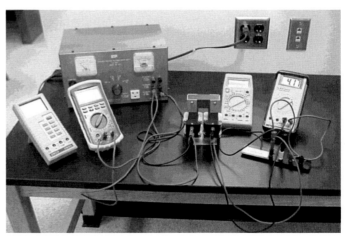

Procedure

1. Identify the primary and secondary coils of the transformer for step-up mode. (The smaller coil should be at the primary side.)

2. Connect the transformer circuits as shown below, with AC voltmeters connected in parallel at the coils and AC ammeters connected in series. Switches are optional.

3. Turn on the **AC power** (not DC!) source. Close switches. Adjust the **AC power** source to supply 6 volts to the primary coil.
   Caution! The load resistor can get very hot! Avoid burns!

4. Record input and output voltages and currents in the data table.

5. Increase the AC power source to 10 volts.

6. Record input and output voltages and currents in the data table.

7. Repeat steps 2 – 6 above but with the transformer in step-down mode. (The larger coil should be at the primary side.)

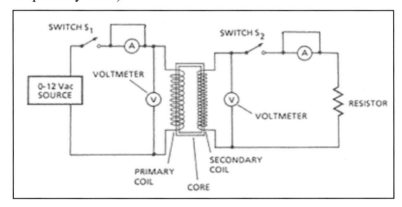

Data Table

	E_p	I_p	E_s	I_s	P_p	P_s	η
Step-up 6V							
Step-up 10V							
Step-down 6V							
Step-down 10V							

Questions

1. In terms of power, what does an ideal electrical transformer do?

2. In terms of voltage, what does a transformer do?

3. In an ideal transformer, when the voltage is increased at the secondary, what change can be expected in the current? When the voltage is decreased at the secondary, what change can be expected in the current?

4. What is the relationship between the primary and secondary *voltages* in this particular transformer when in step-up and step-down mode?

5. In this particular transformer, what is the ratio of the number of *windings* in the primary and secondary coils?

6. What determines whether a transformer is connected in step-up or step-down mode?

AIR GEAR MOTOR **L23**

Student Names(s) _____

_____

_____

Power & Efficiency: Air Gear-Motor

Objectives:

- Describe an air motor. Explain what it does and how it works.
- Use an air motor to lift a load.
- Measure pressure drop across, and flow rate through, an motor.
- Determine the efficiency of an air motor.

Discussion:

An air motor uses compressed air to turn a shaft, converting fluid power into rotating mechanical power. Air motors are often used in place of electrical motors where there is too much moisture and where there are explosive atmospheres – where an electrical spark might be dangerous.

Air motors operate on fluid power, which is the product of the air pressure p and the flow rate $\dot{V}$ given as

$$P_{in} = p\dot{V}$$

Here, the air pressure is measured with the gage on the pressure regular and the flow rate is measure by the **rotameter**.

The output power is translational mechanical or

$$P_{out} = \frac{Fd}{t}.$$

The efficiency of the system is the ratio of the output power to the input power or

$$\eta = \frac{P_o}{P_i}.$$

Equipment:

- Compressed air supply, either compressor or air tank
- Air motor
- Take-up drum, string
- Plastic tubing
- Clamps
- Pressure regulator w/ compound pressure gage
- Rotameter (air flow meter)
- Weight set and hanger
- Stop watch

Procedure:

1. Set up the apparatus as shown in the above figure. Air flows through the regulator, then through the air gear-motor, and finally through to the rotameter air-flow meter. Start with the pressure regulator closed. (Initial tests might be performed here, checking for leaks and adjusting the regulator to check air motor operation.)

2. Hang weight on a string attached to the take-up drum (or pulley). Start with at least three kilos. Add more weight for the second trial, say six kilos.

3. The work done on the weight must be done with the weight moving upward at constant speed. Set a lift distance (about a meter or less) over which the time will be measured. The idea is to match pressure to weight such that a uniform lift speed is achieved, one that can be reasonably measured with a stop watch. Gearing slows the movement down, increasing torque, but reducing efficiency.

4. In two trials, lift two different weights, recording the weight, flow rate, pressure, and time.

Data and Conversion of Units:

Convert volume flow rate ($\dot{V}$) from cfh to ft³/sec by dividing by 3600 sec/hr.

Trial 1: $\dot{V}$ = _____ cfh = _____ ft³/s

Trial 2: $\dot{V}$ = _____ cfh = _____ ft³/s

Convert weight (F) from kg to pounds by multiplying by 2.20 lb/kg.

Trial 1: F = _____ kg = _____ lb

Trial 2: F = _____ kg = _____ lb

Convert pressure reading (p) from psi to lb/ft² by multiplying by 144 in²/ft².

Trial 1: p = _____ psi = _____ lb/ft²

Trial 2: p = _____ psi = _____ lb/ft²

Power and Efficiency Calculations

- Calculate fluid input power $P_{in} = p\dot{V}$ and record in the table below.
- Calculate mechanical power $P_{out} = \dfrac{Fd}{t}$ and record in the table below.
- Calculate efficiency $\eta = \dfrac{P_{out}}{P_{in}}$ in percent and record in the table below.

Data and Calculations Table

Trial	d (ft)	F (lb)	t (sec)	p (lb/ft²)	$\dot{V}$ (ft³/s)	P_{out} (ft-lb/s)	P_{in} (ft-lb/s)	η
1								
2								

Wrap-up

1. In terms of power, explain what an air motor does.

2. What was the effect on the pressure and flow rate when the motor was required to raise a heavier load?

3. Why was it necessary to move the weight at constant speed as possible rather than allowing it to accelerate?

4. Why was it necessary to convert the psi and cfh units?

5. Describe a situation where an air motor would be a better choice over an electric motor.

MOTOR – GENERATOR **L24**

Student Names(s) _____

_____

_____

Power & Efficiency: Motor-Generator

Objectives

- Determine the power supplied to an electric motor.
- Determine the output power of a generator.
- Determine the efficiency of a motor-generator system.

Proficiencies

- Set up small motor-generators to generate electrical power using a DC motor as an energy source.
- Measure power using voltmeters and ammeters.
- Calculate the power consumed by an electric motor.
- Calculate the electrical power output of a DC generator.
- Calculate power efficiency.

Discussion

Electric motors and generators are devices that convert power from one form to another. Motors convert electrical power into rotating mechanical power. Generators convert rotating mechanical power into electrical power. The main parts of motors and generators are very similar, one designed to operate more efficiently in either particular mode of operation. Generally, they differ only in how they are used.

The power input in watts (P_{in}) to the system is the product of the input voltage (in volts) and input current (in amps) or

$$P_{in} = E_{in}I_{in}$$

Similarly, the power output in watts (P_{out}) to the system is the product of the output voltage (in volts) and output current (in amps) or

$$P_{out} = E_{out}I_{out}$$

The voltage is measured by selecting DC volts on the multi-meter and simply connecting the meter *in parallel* across the load. The current is measure by selecting DC amps on the multi-meter and connecting the meter *in series* with the load.

The efficiency of the motor-generator system is the ratio of the output power to the input power or

$$\eta = \frac{P_{out}}{P_{in}}$$

Equipment

- Allen wrench
- Two DC motors, one operating as a generator
- Four multi-meters (voltage and current at the input and the output)
- Shaft "Lovejoy" coupler or vinyl tube to mechanically connect shafts
- Lead set
- DC power supply
- 6 - 12 volt lamps w/sockets

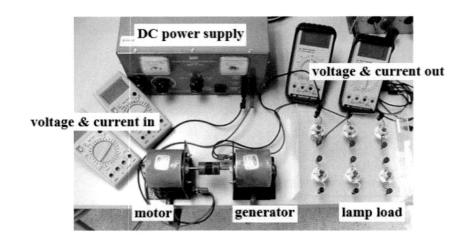

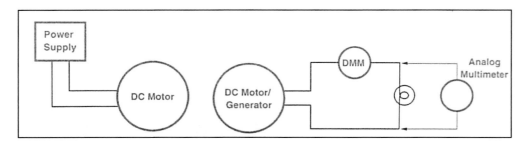

Procedure

1. Couple the two motor shafts together with the coupler, securely clamping the motor bases to the lab bench. Adjust the alignment so that the linkage is smooth and even.

2. With the power supply OFF, connect the DC power supply (0-24 volts DC) to one of the motors using two universal leads. This one serves as the motor.

3. The other motor serves as the generator. Using universal leads, connect the generator to the lamp load.

4. As this point a test run can be done, rotating the system such that the lamp lights up.

5. With the leads in the proper meter sockets and the meter in DC volts mode, connect the input voltmeter across the motor.

6. With the leads in the proper meter sockets (DC volts) and the meter in DC volts mode, connect the output voltmeter across the lamp.

7. With the leads in the proper meter sockets (DC volts) and the meter in DC amps mode, break the circuit to the motor and insert the ammeter to read input current.

8. With the leads in the proper meter sockets (DC amps) and the meter in DC amps mode, break the circuit to the motor and insert the ammeter to read input current.

9. With the lead in the proper meter sockets (DC amps) and the meter in DC amps mode, break the circuit to the lamp and insert the ammeter to read output current.

10. Turn on the power supply and adjust the voltage to 6VDC or such that the lamp is slightly illuminated. Do not exceed the motor rated 12 volt limit.

11. Record the voltages and currents in the table.

12. Calculate and record the input power.

13. Calculate and record the output power.

14. Calculate and record the system efficiency.

15. Repeat steps 10-14 for 9 volts and for 12 volts. Theoretically, efficiency should increase as the system reaches design-rated 12 volts.

Data and Calculations Table

	Trial 1	Trial 2	Trial 3
E_{in}			
I_{in}			
P_{in}			
E_{out}			
I_{out}			
P_{out}			
η			

Questions and Interpretations: Motor − Generator

1. Electric motors transform _____ power into _____ power.

2. Electric generators transform _____ power into _____ power.

3. At which voltage did the motor/generator combination have the highest efficiency?

4. Student Challenge: Some microwave ovens are rated at 900 watts. If 110 volts are supplied to the oven, how much amperage is used by the oven?

PHOTOVOLTAIC PANELS L25

Student Name(s) _____

_____

_____

Power & Efficiency
Photovoltaic Panels

Purpose

The student will study the efficiency of photovoltaic panels in converting radiant energy to electrical energy.

Discussion

Photovoltaic materials are semiconductor devices that convert light energy into electrical energy. When exposed to light, the junctions of some semiconductor materials deposited on a metal plate generate a voltage. The absorption of light causes a measurable voltage difference across the electrical contacts (or protruding wires). When an external circuit is connected, a small current can be drawn from the device. The metal-semiconductor "sandwich" thus behaves like a small battery as long as light is received. This conversion of light energy to a voltage difference is called the **photovoltaic effect**.

In this exercise the photovoltaic cells studied are small solar panels. They convert radiant energy to electrical energy. Solar cells typically have efficiencies less than 10%. The efficiency is defined, as usual, by the ratio of power out to power in. So to determine the efficiency of a solar panel, one needs to know the radiation power incident on the panel. This can be determined by using a "photometer" calibrated in watts of radiant power.

Photometers

Two different photometers are used in this exercise. Students are encouraged to try both and compare results. Each of these devices measures radiant energy in milliwatts per active area. The "active area" is the small hole (with sometimes a lens or glass cover) that limits the light striking the internal detector. As a result, measurements made with these instruments are in milliwatts per active area. The Metrologic brand photometer (the white one) has an adjustable active area cap. with $0.020\,cm^2$ is used here.

The Industrial Fiber Optics brand photometer (the blue one) has a larger active area of approximately $0.38\,cm^2$.

Divide the instrument wattage measurement by its active area to get a power per area rate (milliwatts per cm^2), then multiply this value by the panel area in cm^2. To calculate efficiency, this radiant input power measurement can then be compared to the electrical output power produced by the panel.

Materials

- Two small solar panels
- Two digital multi-meters, one to measure voltage, the other milliamps
- Photometer
- Flood lamp assembly (or any other light source such as a reading lamp or direct sunlight)
- Resistor (one between 100 to 1000 Ohm to serve as load)
- Connectors (either clip leads or banana plugs)

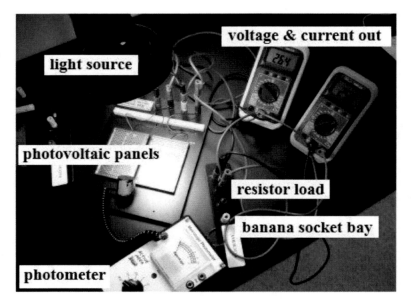

light source

voltage & current out

photovoltaic panels

resistor load

banana socket bay

photometer

Procedure

1. Set up the flood lamp. Adjust the position of the lamp so that the light covers the solar panels relatively uniformly as shown. Do not overheat the panels by placing the lamp too close.

2. Measure the length and width of a solar panel.

3. Turn on the lamp and measure the radiant power at the center and at all four corners of each solar-cell panel with the photometer. Average these readings to get an average incident power on the panels. This measurement is the total of the room light and lamp light. Turn the lamp off.

Part A: One Solar-Cell Panel

1. Without moving the solar panels or the lamp, connect a solar-cell as shown, with the resistor serving as the load, the voltmeter connected in parallel to the load, and the ammeter connected in series with the load.

2. Turn the lamp on. Record the output current and voltage at the load.

3. Turn the lamp off.

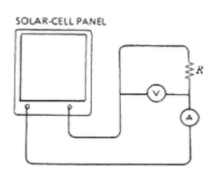

Part B: Series-Connected Solar-Cell Panels

1. Without moving the solar panels or the lamp, connect two solar-cell panels in series as shown.

2. Turn the lamp on. Record the output voltage and current.

3. Turn the lamp off.

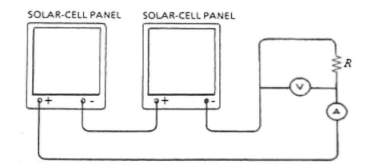

Part C: Parallel-Connected Solar-Cell Panels

1. Without moving the solar panels or the lamp, connect two solar-cell panels in parallel as shown.

2. Turn the lamp on. Record the output voltage and current.

3. Turn the lamp off.

.

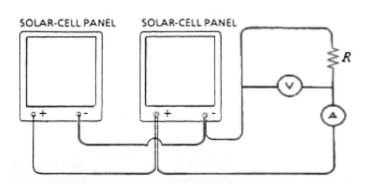

DATA TABLES

Input Power

Panel dimensions: width = _____ cm, length = _____ cm,   area = _____ cm^2

Average photometer power reading over single solar panel = _____ mW per active area.

Single Panel Input Power = (photometer reading ÷ active area)(panel area)

= (_____ ÷ active area)(_____) = _____ mW

Two Panel Configurations = 2(single panel power) = _____ mW

Output Power	Output Voltage	Output Current	Output Power P = EI
Single Panel			
Series Connected			
Parallel Connected			

Efficiency	Output Power	Input Power	Efficiency $\eta = \dfrac{P_o}{P_i}$
Single Panel			
Series Connected			
Parallel Connected			

Questions – Photo Cells

1. How many solar-cell panels rated at 18 Volts and 100 mA would be needed to supply 1 kW of power? Should these be connected in series or parallel?

2. Is there a significant difference in the efficiency of the solar panels when connected in either series or parallel? Which configuration provided the most power?

3. Optional: Whether your cells worked better while connected in series or parallel depends largely on the value of load resistor. Try repeating the lab with a resistor value different than the one you first selected, a value on the other end of the recommended range, to determine how this effects the efficiency when connected in parallel or series.

4. What factors would you consider if you were to determine the location of a solar-cell power source?

5. What is the output power of a solar panel under no-load conditions? (Hint: How much current flows under no-load?)

HEAT EXCHANGER **L26**

Student Name(s) _____

_____

_____

Power & Efficiency

Heat Exchanger

Purpose

The student will experimentally determine the power of a heat exchanger.

Discussion

We normally associate power with a source of energy. This is true for heaters, devices that provide thermal power or thermal energy. For most heat exchangers, however, the goal is to *remove* heat energy.

In this lab, you will determine the thermal power of a heat exchanger designed to remove heat. By removing heat from $100°C$ steam, you will condense it to distilled water, and then cool it still further to less than $100°C$. A powerful heat exchanger can do this very rapidly. The rate at which the particular exchanger in this lab is able to condense and further cool is the measure of the thermal power of the system.

Heat energy that is added to or removed from a substance can have two effects – a change of state and/or a change in temperature. A change of state means a conversion from a solid to liquid or from a liquid to vapor. Heat that causes a change of state is called **latent heat**. The heat energy required to change one gram of water from boiling point to steam at the same temperature is called its **heat of vaporization**, abbreviated H_v. The latent heat energy required to covert boiling water to steam, or visa-versa, is given by

$$Q = mH_v$$

where *m* is the mass in grams and H_v is 539 calories/gram for water.

When heating or cooling does not alter the state of a substance, heating or cooling simply changes its temperature. Heat that causes changes in temperature is called **sensible heat**. The amount of heat required to change the temperature of one gram of a substance by one Celsius degree is called its **specific heat**, c. The amount of sensible heat required to change the temperature of m grams of a substance by ΔT degrees Celsius is given by

$$Q = mc\Delta T$$

where c has a value of 1 calorie/gram $\cdot\ C°$ for water.

The unifying principle for power, energy divided by time, as applied to this process is the total heat energy displaced divided by time or

$$P = \frac{Q}{t} = \dot{Q}$$

The thermal power absorbed or liberated for the heat exchanger is

$$Power = \frac{calories\ removed\ by\ heat\ exchanger}{time\ to\ remove\ heat}$$

Both latent and sensible heat energy are liberated…

$$Power = \frac{calories\ to\ condense\ steam + calories\ to\ cool\ water}{time\ to\ collect\ condensate}$$

Substituting the formulas for latent and sensible heat as applied to water we have

$$Power = \frac{mH_v + mc\Delta T}{t}$$

$$Power = \frac{(m)(539 cal\,/\,g) + (m)(1\,cal\,/\,g\cdot C°)(100°C - T_f)}{t}$$

For an easy way to convert the power from calories per second to watts (J/sec), multiply by 4.1861 J/cal.

Equipment

Heat exchanger apparatus:
    Radiator
    Fan
    Tubing
Thermocouple w/meter
Boiler
Hot plate
Beaker
DC power supply
Timer

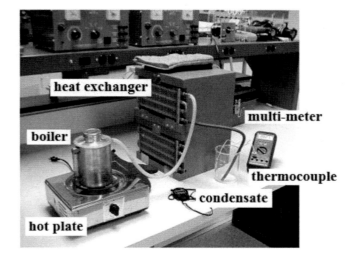

Procedure

1. Set up the apparatus as shown. The primary concern is to route the steam safely from the boiler unit to the exchanger. Be sure the tubing is secure and that steam can travel freely through the exchanger to the outlet port.

2. When the setup has been approved by your instructor, turn the hot plate to maximum and bring the water to a steady boil.

Caution: Steam can cause serious burns. Be very careful with the tubing carrying the steam.

3. Connect the exchanger fan to the DC power supply. Adjust it to 12 VDC.

4. Position the thermocouple junction in the stream of condensate coming out of the heat exchanger. Run the steam through the heat exchanger until the condensate at the outlet reaches a steady and constant temperature.

Record this final temperature: $T_f =$ _____ $°C$.

5. Empty any condensate that may have collected in the beaker. Replace the beaker under the condensate stream and start the timer. Collect 100 to 200 grams of condensate (remember since the density of water is 1 g/cc, 100 mL water equals 100 cc equals 100 grams). Record this value in the table. When the collection is complete, stop the timer and shut everything off. (m = 100 to 200grams)

6. Apply the power equation given in the discussion above to calculate the power in calories per second. Write the entire equation with the numerical substitutions.

P = _____ cal/sec

7. Use the conversion factor given in the discussion above to convert the power from cal/sec to Watts.

P = _____ Watts

Questions – Heat Exchanger

1. In terms of power, explain what a heat exchanger does.

2. Heat was removed from the steam, changing it to water. Then the water was cooled further to a lower temperature. Where did this energy go?

3. What effect on the thermal power of this heat exchanger would you expect if you reduced the amount of air flow flowing through the exchanger, say by reducing the power supply voltage of the fan to six volts? How would you verify this?

4. Explain why the room temperature can greatly affect the thermal power of the heat exchanger. How would your results have been different if the room temperature were 20 degrees cooler? 20 degrees warmer

5.6 STUDENT EXERCISES   POWER & EFFICIENCY

1. Explain in words the unifying principle for power.

2. Power is often measured in horsepower, but there are other units.
   a. The standard metric unit for power is _____.
   b. The standard English unit for power is _____.

3. Explain in words who James Watt was as well as the origins of the horsepower measure of power.

4. How much power (in hp) is needed to lift a 500 pound weight to a height of 12 feet in 10 seconds?

5. A 100 Watt light bulb is connected to a 120 volt source. Find the current drawn by the lamp.

6. Convert.
   a. 1650 ft-lb/sec = _____ hp

   b. 3750W = _____ hp

   c. 3500 Btu/hr = _____ W

   d. 5.22 kcal/sec = _____ W

7. An electric teapot is rated 3500 BTU/hr power output. What is the current draw when connected to a 120 volt source? (Assume 100% efficiency.)

8. What's the most amount of torque a 30 hp engine can produce at 600rpm?

9. What is the maximum amount of power available at a 120 volt wall socket on a 15A breaker?

10. A torque of 0.7 ft-lb is supplied at a motor shaft. The motor rotates at an angular speed of 1200 rpm. Find the shaft power in ft-lb/s.

11. A 3000 pound load is to be raised to a height of 50 feet in half a minute. The available electric motors are known to be about 25% efficient. What size electric motor (in horsepower) will do the job?

12. A 2.2 kΩ series resister is connected to a 24-volt source. Calculate the power dissipated by the resistor.

13. Thermal power does not fit the unifying principle. Explain in words how thermal power is defined.

14. Fifty gallons of water must be heated from $70\,^{\circ}F$ to $100\,^{\circ}F$ in twenty minutes. Find the required power in

    a. Btu/hr

    b. Watts .

15. Two hundred thousand gallons of water must be pumped to a height of 75 feet in four hours by a pump known to be twenty percent efficient. Calculate

    a. the output power required.

    b. the input power required.

    c. the minimum horsepower of the pump that will satisfy the requirements.

16. a. In terms of **power**, describe what a heat exchanger does.

    b. In terms of **power**, describe what an electric motor does.

    c. In terms of **voltage**, describe what an electric transformer does.

17. An air motor is two percent efficient. It is powered by an air compressor capable of providing 120 cfm at 90 psi. Calculate

    a. the input power.

    b. the output power.

    c. the torque produced at 300 rpm.

18. Radiant energy from the sun varies dramatically depending on weather conditions. Today, according to our photometer reading, the energy available on the photovoltaic panel is 15mW per square centimeter.

   a. How much power will a 10% efficient solar panel measuring 1.5 m by 2.0 m produce?

   b. If the system is regulated at 12 volts, how much current can be expected?

19. How much power must be produced by a pump operating at 50 psi and moving fluid at a rate of 20 gpm?

20. The 45 hp tractor has a power take-off (PTO) that can be shifted to rotate at two speeds, 540 rpm and 1080 rpm.

   a. What change in torque can be expected at the PTO shaft when the operator shifts from 540 rpm to 1080 rpm?

   b. Assuming 100% efficiency, calculate the shaft torque at the two available shaft speeds.

21. A service technician from the elevator company is asked to repair an elevator that is constantly tripping its circuit breaker. The technician uses the following information from the elevator specification plate.

Elevator weight: 2450N
Safe Weight (Elevator +Load): 33517N
Hoist Efficiency: 80%
Elevator Speed: 1.5 m/sec
Voltage: 240 AC
Amperage: 150A
Shaft HP: 48.25

Find

   a. The power needed to operate the elevator while carrying its safe weight at the rated speed.

   b. Whether the motor will supply the power needed to operate the elevator.

   c. Explain why the breaker keeps tripping.

Answers:
(1) power is the rate of doing work or the rate of consuming energy
(2a) Watt **(2b)** ft-lb/sec (3) He improved on the stema engine so that it can reliably be used to power ships, locomotives, electric generators, tractors, etc. He also to the horsepower unit.
(4) 1.10 hp **(5)** 0.833A **(6a)** 3.00 hp **(6b)** 5.03 hp **(6c)** 1030W
(6d) 21,900W **(7)** 8.55A **(8)** 263 ft-lb **(9)** 1800W **(10)** 87.9 ft-lb/s
(11) 36.4 hp **(12)** 262mW **(14a)** 37,500 Btu/hr **(14b)** 11.0 kW
(15a) 8690 ft-lb/s **(15b)** 43,400 ft-lb/s **(15c)** 79.0 hp **(17a)** 25,900 ft-lb/s **(17b)** 518 ft-lb/s **(17c)** 16.5 ft-lb **(18a)** 45.0W **(18b)** 3.75A
(19) 320 ft-lb/s **(20a)** 219 ft-lb less at higher speed (half) **(20b)** 438 ft-lb @ 540 rpm, 219 ft-lb @ 1080 rpm **(21a)** 50.3 kW **(21b)** No
(21c) Power demand exceeds available electrical input power of 36.0 kW

Table of Conversion Factors

Length
1 in = 2.54 cm
1 ft = 12 in = 0.3048 m
1 m = 39.37 in
1 mi = 5280 ft = 1.609 km
1 yd = 3 ft
1 rod = 16.5 ft

Area
$1 \ ft^2 = 144 \ in^2 = 929 \ cm^2 = 9.29 \times 10^{-2} \ m^2$
$1 \ m^2 = 10,000 \ cm^2 = 10.76 \ ft^2$
$1 \ acre = 43,460 \ ft^2$
$1 \ square \ mile = 640 \ acre = 27,880,000 \ ft^2$

Volume
$1 \ m^3 = 1,000,000 \ cm^3 = 35.31 \ ft^3$
$1 \ ft^3 = 1728 \ in^3 = 28,320 \ cm^3 = 0.02832 \ m^3$
$= 7.48 \ gal$
$1 \ gal = 231 \ in^3 = 4 \ qts = 8 \ pts$
$= 128 \ fluid \ ounces = 3.79 \ L$
$1 \ L = 1000 \ cm^3 = 0.264 \ gal$

Angles
$1 \ radian = 57.3^0$
$1 \ rev = 2\pi \ rad = 360^0$

Charge and Current
$1 \ coulomb = 6.25 \times 10^{18} \ electrons$
1 A = 1 c/s

Mass, Weight and Force
$1 \ lb = 4.448 \ N = 4.448 \times 10^5 \ dyne$
1 kg = 9.81 N = 2.205 lb
1 slug = 32.2 lb

Speed
1 mph = 1.467 ft/s = 1.609 km/hr
$= 0.4470 \ m/s$

Pressure
$1 \ psi = 27.68 \ in \ H_2O = 5.171 \ cm \ H_2O$
$= 6,895 \ Pa = 0.0681 \ atm = 144 lb/ft^2$

Energy
1 Btu = 779 lb-ft = 1055 J = 252.0 cal
$= 2.930 \times 10^{-4} \ kWh$

Power
$1 \ hp = 550 \ \frac{ft-lb}{sec} = 746 \ W = 2545 \ Btu/hr$
$= 178.2 \ cal/sec$

Time
1 day = 24 hrs
1 hr = 60 min
1 min = 60 sec

Density and Specific Gravity of Selected Substances

Solids	Density	Specific Gravity
Gold	19.3 g/cm^3	19.3
Lead	11.3 g/cm^3	11.3
Silver	10.5 g/cm^3	10.5
Copper	8.9 g/cm^3	8.9
Brass	8.6 g/cm^3	8.6
Steel	7.8 g/cm^3	7.8
Aluminum	2.7 g/cm^3	2.7
Balsa Wood	0.3 g/cm^3	0.3
Oak Wood	0.8 g/cm^3	0.8
Liquids		
Mercury	13.6 g/cm^3	13.6
Antifreeze	1.13 g/cm^3	1.13
Seawater	1.04 g/cm^3	1.04
Water	1.0 g/cm^3	1.0
Oil	0.9 g/cm^3	0.9
Alcohol	0.8 g/cm^3	0.8

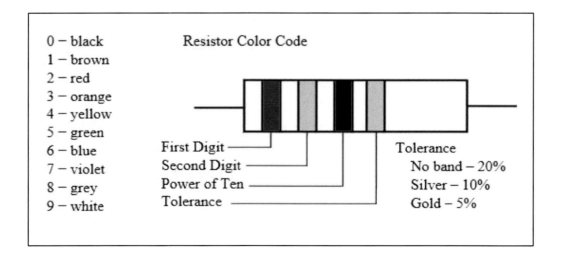

0 – black
1 – brown
2 – red
3 – orange
4 – yellow
5 – green
6 – blue
7 – violet
8 – grey
9 – white

Resistor Color Code

First Digit
Second Digit
Power of Ten
Tolerance

Tolerance
No band – 20%
Silver – 10%
Gold – 5%

TABLE OF TRIGONOMETRIC FUNCTIONS

Angle	Sinθ	Cosθ	Tanθ	Cotθ	
0°	0.0000	1.0000	0.0000	- - -	90°
1°	0.0175	0.9998	0.0175	57.29	89°
2°	0.0349	0.9994	0.0349	28.64	88°
3°	0.0523	0.9986	0.0524	19.08	87°
4°	0.0698	0.9976	0.0699	14.30	86°
5°	0.0872	0.9962	0.0875	11.43	85°
6°	0.1045	0.9945	0.1051	9.514	84°
7°	0.1219	0.9925	0.1228	8.144	83°
8°	0.1392	0.9903	0.1405	7.115	82°
9°	0.1564	0.9877	0.1584	6.314	81°
10°	0.1736	0.9848	0.1763	5.671	80°
11°	0.1908	0.9816	0.1944	5.145	79°
12°	0.2079	0.9781	0.2126	4.705	78°
13°	0.2250	0.9744	0.2309	4.331	77°
14°	0.2419	0.9703	0.2493	4.011	76°
15°	0.2588	0.9659	0.2679	3.732	75°
16°	0.2756	0.9613	0.2867	3.487	74°
17°	0.2924	0.9563	0.3057	3.271	73°
18°	0.3090	0.9511	0.3249	3.078	72°
19°	0.3256	0.9455	0.3443	2.904	71°
20°	0.3420	0.9397	0.3640	2.747	70°
21°	0.3584	0.9336	0.3839	2.605	69°
22°	0.3756	0.9272	0.4040	2.475	68°
23°	0.3907	0.9205	0.4245	2.356	67°
24°	0.4067	0.9135	0.4452	2.246	66°
25°	0.4226	0.9063	0.4663	2.145	65°
26°	0.4384	0.8988	0.4877	2.050	64°
27°	0.4540	0.8910	0.5095	1.963	63°
28°	0.4695	0.8829	0.5317	1.881	62°
29°	0.4848	0.8746	0.5543	1.804	61°
30°	0.5000	0.8660	0.5774	1.732	60°
31°	0.5150	0.8572	0.6009	1.664	59°
32°	0.5299	0.8480	0.6249	1.179	58°
33°	0.5446	0.8387	0.6494	1.540	57°
34°	0.5592	0.8290	0.6745	1.483	56°
35°	0.5736	0.8192	0.7002	1.428	55°
36°	0.5878	0.8090	0.7265	1.376	54°
37°	0.6018	0.7986	0.7536	1.327	53°
38°	0.6157	0.7880	0.7813	1.280	52°
39°	0.6293	0.7771	0.8098	1.235	51°
40°	0.6428	0.7660	0.8391	1.192	50°
41°	0.6561	0.7547	0.8693	1.150	49°
42°	0.6691	0.7431	0.9004	1.111	48°
43°	0.6820	0.7314	0.9325	1.072	47°
44°	0.6947	0.7193	0.9657	1.036	46°
45°	0.7071	0.7071	1.0000	1.000	45°
	Cosθ	Sinθ	Cotθ	Tanθ	Angle

WATER

$\rho_m = 1 \text{ g/cm}^3 = 1 \text{ g/cc} = 1 \text{ g/mL} = 1000 \text{ kg/m}^3$

$\rho_w = 9810 \text{ N/m}^3 = 62.4 \text{ lb/ft}^3 = 8.34 \text{ lb/gal}$

$c = 1 \text{ Btu/lb-F}^0 = 1 \text{ cal/g-C}^0$

$H_v = 540 \text{ Btu/lb} = 540 \text{ cal/g}$

$H_f = 79.8 \text{ Btu/lb} = 79.8 \text{ cal/g}$

$p = 0.433 \text{ psi/ft} = 0.0361 \text{ psi/in}$

Selected Metric Prefixes		
Prefix	**Symbol**	**Value**
pico	p	10^{-12}
nano	n	10^{-9}
micro	μ	10^{-6}
milli	m	10^{-3}
centi	c	10^{-2}
kilo	k	10^{3}
mega	M	10^{6}
giga	G	10^{9}
terra	P	10^{12}

Selected Greek Letters and Their Common Usage				
Name	**Upper Case**	**Usage**	**Lower Case**	**Usage**
Alpha	A		α	angular acceleration
Beta	B		β	angle
Delta	Δ	difference, change in	δ	angle, increment
Eta	H		η	efficiency
Theta	Θ		θ	angular displacement
Iota	I	electric current	ι	
Lambda	Λ		λ	wavelength
Mu	M		μ	prefix micro 10^{-6}
Pi	Π		π	constant c/d = 3.14
Rho	P		ρ	density
Sigma	Σ	summation	σ	
Tau	T		τ	torque
Omega	Ω	resistance unit Ohms	ω	angular velocity

Type K Thermocouple Reference Table

Nickel-Chromium vs. Nickel-Aluminum

Temperature in °F
Reference Junction at 32°F

°F	0°	1°	2°	3°	4°	5°	6°	7°	8°	9°	10°	°F
30	-0.044	-0.022	0.000	0.022	0.044	0.066	0.088	0.110	0.132	0.154	0.176	30
40	0.176	0.198	0.220	0.242	0.264	0.286	0.308	0.330	0.353	0.375	0.397	40
50	0.397	0.419	0.441	0.463	0.486	0.508	0.530	0.552	0.575	0.597	0.619	50
60	0.619	0.642	0.664	0.686	0.709	0.731	0.753	0.776	0.798	0.821	0.843	60
70	0.843	0.865	0.888	0.910	0.933	0.955	0.978	1.000	1.023	1.045	1.068	70
80	1.068	1.090	1.113	1.136	1.158	1.181	1.203	1.226	1.249	1.271	1.294	80
90	1.294	1.316	1.339	1.362	1.384	1.407	1.430	1.453	1.475	1.498	1.521	90
100	1.521	1.543	1.566	1.589	1.612	1.635	1.657	1.680	1.703	1.726	1.749	100
110	1.749	1.771	1.794	1.817	1.840	1.863	1.886	1.909	1.931	1.954	1.977	110
120	1.977	2.000	2.023	2.046	2.069	2.092	2.115	2.138	2.161	2.184	2.207	120
130	2.207	2.230	2.253	2.276	2.298	2.321	2.344	2.367	2.390	2.413	2.436	130
140	2.436	2.459	2.483	2.506	2.529	2.552	2.575	2.598	2.621	2.644	2.667	140
150	2.667	2.690	2.713	2.736	2.759	2.782	2.805	2.828	2.851	2.874	2.897	150
160	2.897	2.920	2.944	2.967	2.990	3.013	3.036	3.059	3.082	3.105	3.128	160
170	3.128	3.151	3.174	3.197	3.220	3.244	3.267	3.290	3.313	3.336	3.359	170
180	3.359	3.382	3.405	3.428	3.451	3.474	3.497	3.520	3.544	3.567	3.590	180
190	3.590	3.613	3.636	3.659	3.682	3.705	3.728	3.751	3.774	3.797	3.820	190
200	3.820	3.843	3.866	3.889	3.912	3.935	3.958	3.981	4.004	4.027	4.050	200
210	4.050	4.073	4.096	4.119	4.142	4.165	4.188	4.211	4.234	4.257	4.280	210
220	4.280	4.303	4.326	4.349	4.372	4.395	4.417	4.440	4.463	4.86	4.509	220
230	4.509	4.532	4.555	4.578	4.601	4.623	4.646	4.669	4.692	4.715	4.738	230
°F	0°	1°	2°	3°	4°	5°	6°	7°	8°	9°	10°	°F

absolute (total) pressure The sum of atmospheric pressure and gage pressure. The total pressure in an enclosed volume measured above zero.

alternating current (AC) Electric current in which the direction of electric charge flow changes at a regular rate.

ammeter A device that measures electric current

ampere (A) The unit of measurement for electric current. One ampere equals one coulomb per second.

amplitude The intensity of a wave, usually represented graphically on the vertical scale.

angular acceleration The rate of change of angular speed in a rotating system.

angular displacement The change in angular position traveled when an object rotates from one position to another.

Archimedes' principle An object immersed in a fluid has an upward (buoyant) force equal to the weight of the fluid displaced by the object.

barometer An instrument used for measuring atmospheric pressure.

battery A source of electric potential difference (voltage) in an electric circuit. Converts chemical energy into electric energy. Most common source of DC.

boiling point The temperature at which a material changes state from liquid to gas as thermal is added to the material.

British thermal unit (Btu) Unit of energy in the English system. Defined as the energy required to change the temperature of one pound of water one degree Fahrenheit.

buoyancy (buoyancy force) The upward force exerted on an object immersed in a fluid.

calorie (cal) An SI unit of energy. Defined as the energy required to change the temperature of one gram of water one degree Celsius.

conductor A material containing many free electrons that move through the material easily when an electric field or potential difference is applied.

coulomb (c) The SI unit of electric charge. One coulomb equals 6.25×10^{18} electrons.

current (I) The rate of electron flow in a conductor measure in amperes.

density A property of a material defined as the mass of the material divided by the volume.

direct current (DC) An electric current in which charge flows in one direction in a circuit.

displacement A vector quantity that defines the distance and direction between two points. The magnitude of the displacement vector is the change in position, or distance, and the direction is the angle as measured from a standard starting point.

efficiency The ratio of output work to input work. Can also be define as the ratio of output power to input power.

electrical resistance The measure of the ability of an electrical device to oppose the flow of charge through the device. In terms of Ohm's Law, defined as the voltage drop across the device divided by the current flowing through the device.

electric charge The displacement quantity in electrical systems. A quantity of electrons. The property of producing electromotive force.

electromagnetic induction The process of generating a current in a wire due to the relative motion between the wire and a magnetic field.

electromotive force (EMF) The electrical potential difference produced by either electromagnetic inductance or the chemical processes in a battery. The mover quantity in electrical systems.

energy The property of a systems or object that enables it to do work.

energy dissipation The conversion of work or other useful forms of energy into thermal or other unusable forms of energy. Energy lost.

equilibrium A state in which the net force and/or net torque in a system is zero.

fluid A liquid or gas. A material that can flow, has no definite shape of its own, and conforms to the shape of its container.

foot (ft) The base, fundamental unit of length in the English system.

foot-pound (ft-lb) The English units of torque.

force A push or a pull exerted on an object. Equal to the mass of the objects multiplied to the acceleration of the objects

frequency A measure of how quickly a pattern repeats itself. IN waves and in alternating currents, defined as the ration of the number of complete cycles, or oscillations, of the patter to the time interval over which the cycles are measured.

gage pressure The pressure in an enclosed volume measure above atmospheric pressure.

heat The energy transferred between objects because of temperature differences between the objects.

heat flow rate The speed at which heat is displaced, or the heat displaced divided by the time interval over which the displacement is measured.

heat of fusion The amount of energy required to change the state of a substance from a solid to a liquid or liquid to solid.

heat of vaporization The amount of energy required to change the state of a substance between states of liquid and gas.

hertz (Hz) The unit of measure for frequency. Defined as one cycle per second or one cps.

horsepower (hp) An English unit of power. Define by James Watt as 550 ft-lb/sec.

hydraulic system A system that uses a liquid as the working fluid.

insulator A material that does not contain a significant number of free electrons that can move freely through the material. A material with high electrical resistance.

joule (J) The SI units of energy, equal to one newton-meter.

kilogram (kg) The fundamental unit of mass in the SI system of units.

kilowatt-hour (kWh) The energy consumed by a one kilowatt device operated for one hour.

kinetic energy Energy possessed by an object or system due to the motion of the object or system.

law of conservation of energy The total energy of an isolated system is constant.

lever arm The shortest distance between an object's axis of rotation to the perpendicular ine of   action of an applied force.

mass The measure of the total amount of material contained in an object. The measure of an object's inertia.

mass flow rate The rate at which mass is displaced. The amount of displaced mass divided by the time interval over which the displacement is measured.

melting point The temperature at which a material changes state from solid to liquid as thermal energy is added to the material.

meter (m) The fundamental unit of length in the SI system.

metric system (SI) A set of standards of measurements simplifying fluid volume standards, mass verses weight, and other deficiencies in the English system. Units are conveniently related by powers of ten.

mover The quantity that causes motion or change in an energy system.

newton (N) The fundamental unit of force in the SI system.

Newton's Law $F = ma$. The acceleration of an object is directly proportional to the force exerted on the object but inversely proportional to the mass of the object.

ohm (Ω) The unit of electrical resistance. In terms of Ohm's Law, defined as one volt per ampere.

Ohm's Law Fundamental relationship between the voltage, current, and resistance through an electrical device expressed as $E = IR$.

parallel circuit An electrical circuit where there are multiple paths for current to flow.

Pascal's press Device creating a mechanical advantage, of force multiplier, through the use of two connected cylinders of different size. The ideal mechanical advantage of this system is equal to the ratio of the cylinder areas.

Pascal's principle A change in pressure at any point in a confined fluid is transmitted undiminished throughout the fluid.

period In electromagnetic waves, the time required for one complete cycle.

photoelectric (photovoltaic) effect The liberation of an electric charge by electromagnetic radiation, including solar radiation, striking its surface.

pneumatic system A system that uses a gas as the operating fluid.

potential energy Energy stored by an object or system.

pound (lb) The fundamental unit of force in the English system.

pound-foot (lb-ft) The English unit of work and energy.

power The rate of doing work or the rate of transferring energy.

pressure Force applied over an area, defined by the applied force divided by the area on which it acts.

quadrant One of four equally sized circular sectors, numbered I –IV, of which a full circle is divided.

quantity Any property, characteristic, or effect of the natural world that can be quantified (measured)

radian A dimensionless measure of an angle. Defined as the ratio of arc length of a circular sector divided by the radius.

rate A displacement quantity divided by the interval of time over which the displacement quantity is measured.

resistor An electrical device that has a specific value of resistance.

resultant vector A single vector representing the sum of two or more added vectors.

resultant torque A single value of torque representing the sum of two or more values of torque having the same axis of rotation.

scalar quantity a quantity that can be fully described with magnitude.

second (sec, s) The base unit of time in the SI system.

series circuit An electric circuit in which there is only one path for the current.

slug The unit of mass in the English system.

specific heat The amount of energy that will change the temperature of a unit of mass of the material one degree.

temperature Property of an object proportional to the amount of heat energy in the object. The average kinetic energy of the random motion of the atoms and molecules in the object.

thermal energy The total energy of random motion of vibration of all the particles that make up the object.

thermocouple An electronic device used to measure temperature. Metals of different composition in contact forming a junction will produce a small electrical charge proportional to the junction temperature.

thermometer A device that measures temperature.

torque A quantity that causes rotation in mechanical systems. Defined as the applied force multiplied to the perpendicular lever arm.

vector quantity A quantity that must be described by both magnitude and direction.

volt (V) The standard unit of measurement of electrical potential difference.

voltmeter A device that measures voltage or electrical potential difference.

volume flow rate The change in volume divided by the time interval over which the volume change is measured.

watt (W) The standard SI unit of measurement for power. Defined as one Joule per second.

wavelength The physical length of one cycle of a wave.

weight The force on an object caused by gravity.

weight density Property of a material defined as the weight of the material divided by its volume.

work The amount of energy used in a system to cause a change in the system. Defined as the mover quantity of the energy system multiplied by the displacement, or change, in that energy system, except in thermal systems where the work is the heat energy itself.

References

Principles of Technology, ISBN 1-55502-371-1, *Physics in Context An Integrated Approach,* ISBN 1-57837-275-5, *Physics for Technicians Laboratory Exercises,* ISBN 1-55502-393-2, and *Unified Technical Concepts,* ISBN 1-55502-393-2, Center for Occupational Research and Development (CORD), Waco, Texas

Physics Laboratory Manual, Sanders College Publishing, David Loyd, ISBN 0-03-043904-3

Experiments in College Physics, 9th Edition, Heath and Company , B. Cioffari & D. Edmunds, 6th Edition, D.C., ISBN 0-669-00489-8

Laboratory Manual to Accompany Applied Physics, 8th Edition, Prentice Hall, D. Ewen, N. Schurter, P. Gunderson, S. Rao, ISBN 0-13-110353-9

Experiments for Technical Physics, 2nd Edition, Allyn and Bacon, Inc, Aaron McAlexander, ISBN 0-205-06088-9

Physics Laboratory Manual, Monkey Publishing, J.P. Levasseur, ISBN 9780983297659

Conceptual Physics, 10th Edition, Addison Wesley, Paul G. Hewitt, ISBN 0-8053-9375-7

Applied Physics, 8th Edition, Pearson Publishing, Prentice Hall, D. Ewen, N. Schuter, P.E. Gunderson, ISBN 0-13-110169-2

Basic Technical Mathematics with Calculus, 10th Edition, Pearson Publishing, Addison Wesley, A.J. Washington, ISBN-13 978-0-13-311653-3

Physics, 7th Edition, McGraw Hill, P.E. Tippens, ISBN-13 978-0-07-301267-4

Mathematics for the Trades, 7th Edition, Pearson Publishing, Prentice Hall, R.A. Carman, H.M. Saunders, ISBN 0-13-114525-8

Encyclopedia of Physics, VCH Publishers, Inc., R. G. Lerner, G.L. Trigg, ISBN 0-895573-752-3

Omega Temperature Measurement Handbook, 7th Edition, 2010, Omega Engineering, Inc., (Type K Thermocouple Reference Table)

Wikipedia, the Free Encyclopedia, Scientists' Images and Biographies; Pythagoras, Charles–Augustin de Coulomb, André-Marie Ampère, Isaac Newton, Blaise Pascal, Archimedes, Count Volta, Georg Simon Ohm, James Prescott Joule, James Watt, and Nikola Tesla

Scientific Method in Practice, Cambridge University Press, 2003, Gauch, H.G.

Made in the USA
Columbia, SC
03 September 2020